SUSTAINABLE SOLID WASTE MANAGEMENT

SUSTAINABLE SOLID WASTE MANAGEMENT

Edited By
Syeda Azeem Unnisa, PhD and S. Bhupatthi Rav, PhD

Apple Academic Press

TORONTO   NEW JERSEY

© 2013 by
Apple Academic Press Inc.
3333 Mistwell Crescent
Oakville, ON L6L 0A2
Canada

Apple Academic Press Inc.
1613 Beaver Dam Road, Suite # 104
Point Pleasant, NJ 08742
USA

First issued in paperback 2021

Exclusive worldwide distribution by CRC Press, a Taylor & Francis Group

ISBN 13: 978-1-77463-212-3 (pbk)
ISBN 13: 978-1-926895-24-6 (hbk)

Library of Congress Control Number: 2012935663

Library and Archives Canada Cataloguing in Publication

Sustainable solid waste management/edited by Syeda Azeem Unnisa and S. Bhupatthi Rav.

Includes bibliographical references and index.
ISBN 978-1-926895-24-6
1. Refuse and refuse disposal. 2. Sustainable development. 3. Waste minimization. 4. Recycling (Waste, etc.).
I. Rav, S. Bhupatthi II. Unnisa, Syeda Azeem

TD791.S87 2012   363.72›8   C2011-908740-5

Contents

List of Contributors

Quetzalli Aguilar-Virgen
Facultad de Ingeniería Ensenada, Universidad Autónoma de Baja California, Km 103 Carretera Tijuana-Ensenada, Ensenada, Baja California. México C. P. 22870.

Dolores Elizabeth Turcott Cervantes
Universidad Tecnologica de Leon, Leon, Guanajuato, Mexico.

William Hogland
School of Natural Sciences, Linnaeus University, Landgången 3, SE-391 82 Kalmar, Sweden.

J. Sudhir Kumar
Associate Professor, Dept. of Mechanical engg, Sir C.R.Reddy college of engg, Eluru.

Steffen Lehmann
Zero Waste SA Research Centre for Sustainable Design and Behaviour (sd+b), University of South Australia, Adelaide, SA-5001, Australia.

Mario Bernardo Reyes Marroquín
Universidad Tecnologica de Leon, Leon, Guanajuato, Mexico.

M. A. Memon
International Environmental Technology Centre (IETC), Division of Technology Industry and Technology, United Nations Environment Programme, 2-110 Ryokuchi Koen, Tsurumi-ku, Osaka 538-0036, Japan.

Viatcheslav Moutavtchi
School of Natural Sciences, Linnaeus University, Landgången 3, SE-391 82 Kalmar, Sweden.

Sara Ojeda-Benítez
Instituto de Ingeniería, Universidad Autónoma de Baja California, Blvd. Benito Juárez y Calle de la Norma S/N, 21280 Mexicali B.C. México.

V. I. Otti
Civil Engineering Department, Federal Polytechnic Oko, Anambra State, Nigeria.

Prasada Rao.P. V. V.
Professor, Dept of Environment Science, College of Engineering, Andhra University, Vishakapatnam.

S. Bhupatthi Rav
Research Officer, Regional Centre for Urban and Environmental Studies, Osmania University, Hyderabad, 500007, A.P, India.

Karina Guadalupe López Romo
Universidad Tecnologica de Leon, Leon, Guanajuato, Mexico.

Antonina Shepeleva
Department of Ecological Safety and Sustainable Development of Regions, Faculty of Geography and Geo-ecology, St Petersburg State University, St Petersburg, Russia.

Jan Stenis
School of Natural Sciences, Linnaeus University, Landgången 3, SE-391 82 Kalmar, Sweden.

Venkata Subbiah K.
Professor, Dept of Mechanical Engineering, College of Engineering, Andhra University, Vishakapatnam.

Paul A. Taboada-González
Facultad de Ingeniería Ensenada, Universidad Autónoma de Baja California, Km 103 Carretera Tijuana-Ensenada, Ensenada, Baja California. México C.P. 22870.

Syeda Azeem Unnisa
Research Officer, Regional Centre for Urban and Environmental Studies, Osmania University, Hyderabad-500007, A. P, India.
Director and Dean, Regional Centre for Urban and Environmental Studies, Osmania University, Hyderabad-500007, A. P, India.

Carolina Armijo-de Vega
Facultad de Ingeniería Ensenada, Universidad Autónoma de Baja California, Km 103 Carretera Tijuana-Ensenada, Ensenada, Baja California. México C.P. 22870.

David C. Wyld
Department of Management, Southeastern Louisiana University, Hammond, LA USA.
Director & Dean, Regional Centre for Urban and Environmental Studies, Osmania University, Hyderabad-500007, A.P, India.

A. U. Zaman
Environmental Engineering and Sustainable Infrastructure, School of Architecture and Built Environment, KTH, Sweden.

List of Abbreviations

AMCS	Advanced Manufacturing Control Systems
ANSEPA	Anambra State sanitation and environmental protection agency
UABC	Autonomous University of Baja California
CP	Cleaner production
COSTBUSTER	Company statistical business tool for environmental recovery
CONCYTEG	Consejo de Ciencia y Tecnologia del Estado de Guanajuato
C&D	Construction and demolition
DTIE	Division of Technology, Industry and Economics
ECO-EE	Ecological–economic efficiency
EUROPE	Efficient use of resources for optimal production economy
ELV	End-of-life vehicles
EPA	Environmental Protection Agency
EU	European Union
EPR	Extended producer responsibility
FCA	Full cost accounting
GPS	Global Positioning System
GHG	Greenhouse gas
HDTV	High-Definition Television
IBA	Incinerator bottom ash
IDW	Institut der Deutschen Wirtschaft
ITESI	Instituto Tecnologico Superior de Irapuato
ISWM	Integrated solid waste management
IWMS	Integrated waste management system
IRR	Internal rate of return
IETC	International Environmental Centre
LCA	Life cycle assessment
LCIA	Life cycle impact assessment
MCE	Municipal Corporation of Eluru
MSW	Municipal Solid Waste
MSWM	Municipal solid waste management
NSW	New South Wales
PF	Proportionality factor
RF	Radio frequency
RFID	Radio frequency identification
RA	Recycled aggregate

RDF	Refuse derived fuel
SBC	Secretariat for the Basel Convention
SHGs	Self Help Groups
SWM	Solid waste management
SAT	Sustainability assessment of technologies
SMSWS	Sustainable Municipal Solid Waste Management Systems
UNEP	United Nations Environment Programme
UAM	Universidad Autonoma Metropolitana
UNAM	Universidad Nacional Autonoma de Mexico
UTL	Universidad Tecnologica de Leon
UCC	University Collecting Centre
WAMED	Waste management efficient decision
WTE	Waste-to-energy
WHO	World Health Organization

Preface

Planning for urban solid waste management within the framework of sustainable development raises several intra- and intergenerational issues such as public health and the livelihood of the public. Sustainability of waste management is key to providing an effective service that satisfies the needs of the end users. One pillar of sustainable solid waste management is strategic planning. It links another pillar, that of cost analysis of solid waste options, which also links to useful analytical tools.

This book provides research papers by eminent professors, researchers, scientists, and academicians from all over the globe pertaining to solid waste issues, impacts, latest trends, and technologies in solid waste treatment, site assessment, land filling, storage, handling, transportation, and disposal and waste minimization. In addition, it explores waste site remediation and clean-up technologies, new continuing regulatory, and policy statutes.

The contents highlight about the various treatment technologies followed across the globe, such as

- Sanitary landfill, incineration and gasification-pyrolysis can be studied by SimaPro software based on input-output materials flow and applied for analyzing environmental burden by different impact categories.
- A deterministic model for short- and long-term waste management and management information systems, which determine which type of integrated solid waste management option or program can be used to implement minimized cost and maximized benefit (benefit cost ratio) over a long period of planning. The model can be used by the decision makers in finding the solution to environmental, economical, sanitary, technical, and social goals, through the use of equipment, routine maintenance, and personnel.
- Waste separation and recycling programs for paper and cardboard separation at institutions of higher education through social marketing approaches can be proved to be effective in helping reach the desired change for very different initiatives.
- Integrated solid waste management (ISWM) based on the 3R approach (reduce, reuse, and recycle) can be aimed at optimizing the management of solid waste from all the waste-generating sectors (municipal, construction and demolition, industrial, urban agriculture, and healthcare facilities) and involving all the stakeholders (waste generators, service providers, regulators, government, and community/neighborhoods) to streamline all the stages of waste management, i.e., source separation, collection and transportation, transfer stations and material recovery, treatment and resource recovery, and final disposal.
- RFID (radio frequency identification) is poised to help transform to handle trash, or MSW (municipal solid waste). RFID can be employed in the MSW area to both facilitate the growth of PAYT (pay as you throw) use-based billing

for waste management services and to promote incentive-based recycling programs, both of which aim to reduce the amount of trash entering our landfills.

- A waste management efficient WAMED (waste managements' efficient decision) model for the evaluation of ecological–economic efficiency can serve as an informative support tool for decision making at the corporate, municipal, and regional levels. It encompasses cost–benefit analyses in solid waste management by applying a sustainability-promoting approach that is explicitly related to monetary measures.

This book will prove to be a lasting and invaluable reference source to the policymakers, municipal corporations, researchers, environmentalist for policy implication, management activities in safe, hygienic handling and disposal.

— Syeda Azeem Unnisa, PhD

Introduction

Waste management is the collection, transportation, processing or disposal, and managing and monitoring of waste materials. The term usually relates to materials produced by human activity, and the process is generally undertaken to reduce its effect on health, the environment, or aesthetics. Waste management is a distinct practice from resource recovery, which focuses on delaying the rate of consumption of natural resources. The management of wastes treats all materials as a single class, whether solid, liquid, gaseous or radioactive substances, and tries to reduce the harmful environmental impacts of each through different methods.

Management of municipal solid waste (MSW) is one of the major challenges worldwide. Inadequate collection, recycling or treatment, and uncontrolled disposal of waste in dumps lead to severe hazards, such as health risks and environmental pollution. This situation is especially serious in low- and mid-income countries. Cities, which are hubs of rapid economic development and population growth, generate thousands of tons of MSW that must be managed daily. Low collection coverage, unavailable transport services, and a lack of suitable treatment and disposal facilities are responsible for unsatisfactory solid waste management, leading to water, land, and air pollution, and for putting people and the environment at risk.

WHAT IS SUSTAINABLE WASTE MANAGEMENT?

All goods and products contain raw materials and energy. If they are discarded, we are effectively throwing away valuable natural resources. Waste disposal can also have adverse impacts on local air pollution and greenhouse gas (GHGs) emissions. Sustainable waste management is therefore vital for:

- Conserving valuable natural resources
- Preventing the unnecessary emission of GHGs
- Protecting public health and natural ecosystems.

Sustainable waste management is based on the following waste hierarchy. The measures at the top of the hierarchy are always preferable and should be considered first.

1. Reduce or prevent waste arising: waste minimization initiatives to help businesses and households reduce the amount of waste that they create
2. Reuse waste: reuse waste and thus avoid energy-consuming reprocessing
3. Recycle: reprocess waste for further use
4. Energy recovery: generating energy from waste using a variety of technologies
5. Disposal: put waste in landfill sites.

Waste disposal should be seen as a last resort. Not only does waste disposal mean that valuable resources and energy are being thrown away but also that biodegradable waste in landfill can emit methane, a GHG up to 23 times more potent than carbon dioxide. What's more, landfill space is becoming restricted. Parts of England have only a few years' worth of landfill capacity left. Lastly, landfill sites are an eyesore and a source of local pollution.

TREATMENT TECHNOLOGIES

Traditionally the waste management industry has been slow to adopt new technologies such as RFID (Radio Frequency Identification) tags, GPS and integrated software packages that enable better quality data to be collected without the use of estimation or manual data entry.

- Technologies like RFID tags are now being used to collect data on presentation rates for curb-side pick-ups.
- Benefits of GPS tracking are particularly evident when considering the efficiency of ad hoc pick-ups (such as skip bins or dumpsters) where the collection is done on a consumer request basis.
- Integrated software packages are useful in aggregating this data for use in optimization of operations for waste collection operations.
- Rear vision cameras are commonly used for occupational health and safety reasons, and video recording devices are becoming more widely used, particularly concerning residential services.

EDUCATION AND AWARENESS

Waste management is an area that needs education and awareness for global preservation. A declaration is known as the "Talloires Declaration", which is concerned about the ever-increasing environmental pollution and diminution of natural resources. Education for waste management and pollution is very critical to the perseverance of global health and security of humankind. A number of universities and vocational education institutions are working for the promotion of organizations working for this purpose. A number of supermarkets are today also playing their part in encouraging recycling with the introduction of "reverse vending machines". These machines, when are deposited with used recyclable container, produce refunds from the recycling charges.

Waste is no longer treated as the valueless garbage waste; rather it is considered as a resource in the present time. Resource recovery is one of the prime objectives in sustainable waste management system. Different waste treatment options are available today with different waste management capacities. There is no a single technology that can solve the waste management problem. Integrated waste management systems offer the flexibility of waste treatment option based on different waste fraction, such as plastic, glass, organic waste, or combustible waste. Energy and resource recovery is also important and can be recovered through integrated waste management systems. There are different system analysis tools that are available today for decision makers. Technology or strategy can be analyzed by the environmental, social, or environmental point of view.

— **Syeda Azeem Unnisa, PhD**

1 Comparative Study of Municipal Solid Waste Treatment Technologies Using Life Cycle Assessment Method

*A. U. Zaman

CONTENTS

1.1 INTRODUCTION

The aim of the study is to analyze three different waste treatment technologies by life cycle assessment (LCA) tool. Sanitary landfill, incineration, and gasification–pyrolysis

* Corresponding Author Email: atique@kth.se Tel./Fax: +4673 752 5708

of the waste treatment technologies are studied in SimaPro software based on input-output materials flow. SimaPro software has been applied for analyzing environmental burden by different impact categories. All technologies are favorable to abiotic and ozone layer depletion due to energy recovery from the waste treatment facilities. Sanitary landfill has the significantly lower environmental impact among other thermal treatment while gases are used for fuel with control emission environment. However, sanitary landfill has significant impact on photochemical oxidation, global warming, and acidification. Among thermal technology, pyrolysis–gasification is comparatively more favorable to environment than incineration in global warming, acidification, eutrophication, and eco-toxicity categories. Landfill with energy recovery facilities is environmentally favorable. However, due to large land requirement, difficult emission control system and long time span, restriction on land filling is applying more in the developed countries. Pyrolysis–gasification is more environmental friendly technology than incineration due to higher energy recovery efficiency. The LCA is an effective tool to analyze waste treatment technology based on environmental performances.

Waste is no more treated as the valueless garbage; waste is rather considered as a resource in the present time. Resource recovery is one of the prime objectives in sustainable waste management system. Different waste treatment options are available in the current time with different waste management capacities. There is not a single technology that can solve the waste management problem (Tehrani et al., 2009). Integrated waste management system is commonly applied method in many developed countries. Integrated waste management system offers the flexibility of waste treatment option based on different waste fraction like plastic, glass, organic waste, or combustible waste. Energy and resource recovery is also important and can be recovered through integrated waste management system. There are different system analysis tools (Finnveden and Moberg, 2004) that are available at the present time for the decision makers. Technology or strategy can be analyzed by the environmental, social, or environmental point of view. The LCA is a commonly applied tool to analyze environmental burden for waste management technology, as well as system. In this study, three different municipal solid waste (MSW) management options like pyrolysis–gasification, incineration, and sanitary landfill are analyzed by LCA model using SimaPro software (version 7). In addition, for life cycle inventory analysis, CML 2 (Centre for Environmental Studies, University of Leiden) baseline (2000) method has been used. The study is done primarily to assess three different options and to analyze the environmental burden from the three technologies. Results from the comparative study would be helpful for decision-making processes to evaluate environmental performance of the technologies. However, socio-economic and applicability of the technology are also important for decision and policy making processes which are not considered in this study. Especially, considering land requirement and continuous function-ability of sanitary landfill and other two thermal waste treatment options would have the significant differences which influence decision-making choice while considering MSW treatment options. Different studies have already been done for MSW management options to analyze the benefits and problems associated with the processes. Some of the studies are done by Hallenbeck (1995) Consonni et al. (2005), Liamsanguan and Gheewala (2007), Parizek et al. (2008), Grieco and Poggio (2009),

Psomopoulos et al. (2009), and Stehlik (2009). Integrated waste management system (IWMS) is one of the effective strategies to solve waste management problems. The study has been done in the context of Sweden waste treatment system. However, the data for pyrolysis–gasification of waste has been taken from the United Kingdom's research report due to lack of local data by assuming that both Sweden and UK has similar waste content in municipal solid waste.

1.2 MATERIALS AND METHODS

1.2.1 MSW Treatment Technologies

Integrated waste management options are now been applying in most of the developed countries with resource recycle, recovery, and energy generation facilities from the solid waste. Waste-to-energy (WTE) conversion is now considered as one of the optimal methods to solve the waste management problem in a sustainable way. Different mechanical, biological, and thermo-chemical waste-to-energy technologies are now applying for managing MSW.

In this study, three different MSW technologies are analyzed, and those are:
(1) sanitary landfill,
(2) incineration, and
(3) Pyrolysis/gasification

Brief descriptions of these three technologies are given bellow.

1.2.1.1 Landfill

"A landfill is a facility in which solid wastes are disposed in a manner which limits their impact on the environment. Landfills consist of a complex system of interrelated components and sub-systems that act together to break down and stabilize disposed wastes over time" (FCM, 2004). Landfill is very old but still one of the extensively used technologies for MWS management. Most of the landfill does not have the energy production facilities. In this study, a sanitary landfill with energy recovery system has been studied. Landfill gas are generated from the landfill site in different gas generation phases. Generally, five different phases like initial adjustment, transition phase, acid phase, methane fermentation, and maturation phases are observed in waste landfill (Adapted from Farquhar and Rovers; 1973, Pohland, 1987, 1991). A typical WTE generation by landfill process has shown in Figure 1.1

1.2.1.2 Incineration

Incineration is a thermal waste treatment process where raw or unprocessed waste can be used as feedstock. The incineration process takes place in the presence of sufficient quantity of air to oxidize the feedstock (fuel). Waste is combusted in the temperature of 850 °C and in this stage waste converted to carbon dioxide, water, and non-combustible materials with solid residue state called incinerator bottom ash (IBA) that always contains a small amount of residual carbon (DEFRA, 2007).

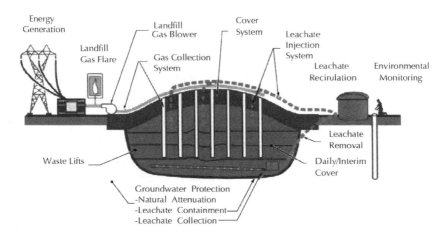

FIGURE 1.1   Principal technical elements of a landfill (FCM, 2004).

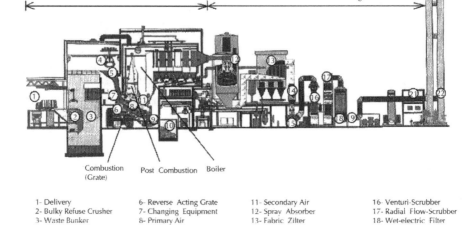

1- Delivery	6- Reverse Acting Grate	11- Secondary Air	16- Venturi-Scrubber
2- Bulky Refuse Crusher	7- Changing Equipment	12- Spray Absorber	17- Radial Flow-Scrubber
3- Waste Bunker	8- Primary Air	13- Fabric Zilter	18- Wet-electric Filter
4- Grab Crane	9- Ash Discharge (Wet)	14- Sound Absorber	19- Clean Gas Reheating
5- Changing Hopper	10- Ash Bunker	15- Induced Draught-Boiler	20- Analytical Room
			21- Chimney

FIGURE 1.2   A schematic MSW combustion plant (Ludwing et al., 2002).

Figure 1.2 shows the schematic diagram of MSW combustion plant where wastes are delivered as feed stock to the pre-combustion (grate) and during post combustion, gas and slug or ashes are produced. Then, in the next phases flue gas is cleaned by water absorber or different filtering methods. Finally, the clean gas is emitted through the chimney to the air. Thermal conversation of waste to energy is now very much applied technology for waste management system due to the generation of heat and energy from the waste stream.

1.2.1.3 Pyrolysis–gasification

Pyrolysis is the thermal degradation of waste in the absence of air to produce gas (often termed syngas), liquid (pyrolysis oil), or solid (char, mainly ash and carbon). Pyrolysis generally takes place between 400–1,000°C. Gasification takes place at higher temperatures than pyrolysis (1,000–1,400°C) in a controlled amount of oxygen (NSCA, 2002). The gaseous product contains CO_2, CO, H_2, CH_4, H_2O, trace amounts of higher hydrocarbons (Bridgwater, 1994). The MSW pyrolysis and in particular gasification is obviously very attractive to reduce and avoid corrosion and emissions by retaining alkali and heavy metals (Malkow, 2004). There would be a net reduction in the emission of the sulphur di-oxide and particulates from the pyrolysis/gasification processes. However, the emission of oxides of nitrogen, VOCs, and dioxins might be similar with the other thermal waste treatment technology (DEFRA, 2004). Figure 1.3 shows the typical flow diagram of the pyrolysis–gasification processes.

1.2.1.4 Life Cycle Assessment

Life cycle environmental assessment tool is one of the effective and principal decision support tools (Christensen et al., 2007) to assess the flow dynamics of the resources. The LCA can give us the idea on environmental burdens per functional unit (kg/ton) of waste generated (Ekvall et al., 2007). Many research works have already been done on LCA all over the world as a decision making tool (Gheewala and Liamsanguan, 2008) for assessing (Bilitewski and Winkler, 2007) waste technology (Ekvall and Finnveden, 2000) models (Björklund, 2000), (Diaz and Warith, 2006) methods (Matsuto, 2002) and strategies (Barton and Patel, 1996; Björklund and Finnveden, 2007; Cherubini et al., 2008; Pennington and Koneczny, 2007) for MSW management. All these study have analyzed waste management options through life cycle perspectives. This study has been done by considering inflow, outflow data, emissions, and resource recovery through electricity and heat recovery from the system. The study is analyzed three different waste treatment technologies that can manage all type of waste fraction.

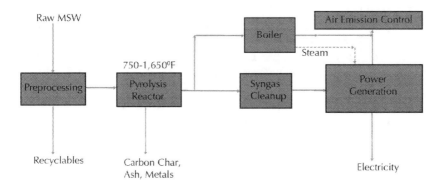

FIGURE 1.3 Typical pyrolysis/gasification system of MSW (Halton EFW Business Case, 2007).

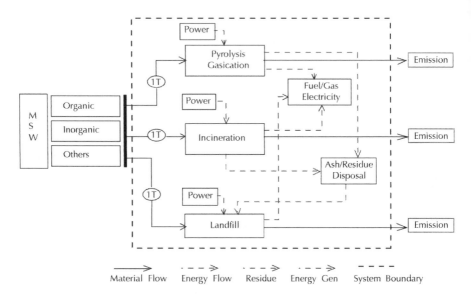

FIGURE 1.4   System boundary for different MSW treatment processes.

1.2.1.5   Aim and Scope

Goal of the study is to develop a LCA model and compare three different MSW treatment options. The study has been carried out by SimaPro (7.0 version) software, life cycle impact assessment has been done by considering CML 2 baseline (2000) method. Waste management technologies are analyzed by ten different impact categories like abiotic depletion, acidification, eutrophication, global warming, ozone layer depletion, human toxicity, fresh water ecotoxicity, marine ecotoxicity, terrestrial ecotoxicity, and photochemical oxidation. Functional unit of the study has been set as one ton of waste mass. Thus, all input and output flows in the model are considered as a reference flow of one ton of MSW treatment for WTE generation. A comparative LCA study has been done in this study. Therefore, average country mix (Sweden) data have been considered for the LCA model while allocating avoiding product. Allocations of the resources have been done based on the system expansion. Figure 1.4 shows the system boundary of the WTE options. Waste is considered as a mixture of compostable or organic, inorganic, and other types of waste fractions.

Within the system boundary, all inputs to the system like 1 ton of MSW and energy requirement for the processes and all outputs like emission to the air waster or soil and final disposal and electricity generation from the systems have been considered.

1.2.1.6   Assumptions

Following assumptions have been made for the LCA model:

- Transport distance of waste for all processes system assumed as same and that is why transportation has been omitted from the system boundary;
- Electricity that produced in the processes is avoided as the average Swedish national electricity production.

1.2.1.7 Life Cycle Inventory and Data Analysis

Life cycle inventory of the LCA model has been made primarily based on the literature, report, and publications. Important papers are Bridgwater (1994), NSCA (2002), Feo et al., (2003), DEFRA (2004), Halton EFW Business Case (2007), Circeo (2009), Khoo (2009). Data emission from the WTE system is shown in the following Table 1.1

TABLE 1.1   Input-output (energy and residue) in different MSW treatment processes

Input/output	Pyrolysis–Gasification	Incineration	Landfill
Start-up energy (kWh/T)	339.3 (3)	77.8 (1)	14.3 (1)+(5)*
Energy generated(kWh/T)	685 (4)	544 (4)	217.3 (1)+(2)
Solid residue (kg/T)	120 (2)	180 (2)	---

Sources: (1) Finnveden et al., (2000), (2) DEFRA (2004), (3) Khoo (2009), (4) Circeo (2009), (5) Cherubini et al. (2008), *Diesel fuel normalized to the energy unit kWh/ton.

In LCA model of Pyrolysis–gasification, the input data have taken as resource (one ton MSW), energy (electricity kWh/ton of MSW), emission (gm/T) to air, soil or waster, energy generation (kWh/ton of MSW) and final residue (kg/ton) produced by the facilities. Table 1.2 shows the emission rate emitted by the facilities during treated one ton of MSW.

TABLE 1.2   Emissions to air from waste management facilities (grams per ton of MSW)

	Emissions to the air from different treatment processes		
Substance	Pyrolysis–Gasification (gm/T)	Incineration (gm/T)	Landfill (gm/T)
Nitrogen oxides	780	1600	680
Particulates	12	38	5,3
Sulphur dioxide	52	42	53
Hydrogen chloride	32	58	3
Hydrogen fluoride	0.34	1	3
VOCs	11	8	6,4
1,1-Dichloroethane	Not likely to be emitted	Not likely to be emitted	0,66
Chloroethane	Not likely to be emitted	Not likely to be emitted	0,26
Chloroethene	Not likely to be emitted	Not likely to be emitted	0,28
Chlorobenzene	Not likely to be emitted	Not likely to be emitted	0,59
Tetrachloroethene	Not likely to be emitted	Not likely to be emitted	0, 98
Benzene	Not likely to be emitted	Not likely to be emitted	0,00006
Methane	Not likely to be emitted	Not likely to be emitted	20,000
Cadmium	0.0069	0.005	0,071
Nickel	0.040	0.05	0,0095
Arsenic	0.060	0.005	0,0012
Mercury	0.069	0.05	0,0012
Dioxins and furans	$4,8 \times 10^{-8}$	$4,0 \times 10^{-7}$	1.4×10^{-7}
Polychlorinated biphenyls	No data	0.0001	No data
Carbon dioxide	10,00,000*	10,00,000	3,00,000
Carbon monoxide	100	No data	---

Source: DEFRA (2004), *CO_2 assumed same as incineration due to same carbon content

Since, carbon content in waste is constant, therefore, for P-G process carbon dioxide emission was assumed same as incineration of municipal solid waste. Model however, developed based on the fossil carbon content (39.5%) in the total carbon emission.

Table 1.3 shows the water emission from the landfill and here surface water and ground water emission are considering as total waster emission.

TABLE 1.3   Emission to the waste from the landfill treatment process

Substances	Emission to water (surface and ground) from landfill (gm/T)*
Aniline	0.00000262
Chloride	30
Cyanide	0.0013
Fluoride	0.0164
Nitrogen (Total)	9.4
Phenols	0.0000077
Phosphorus	0.076
Toluene	0.00019
Arsenic	0.000061
Chromium	0.0009
Copper	0.00014
Lead	0.0012
Nickel	0.0012
Zinc	0.00109

Source: **DEFRA** (2004), * Total emission of water has been counted by adding up the surface and groundwater emission.

1.2.1.8   Life Cycle Impact Assessment (LCIA)

Life cycle impact assessment of the WTE technologies has been done the CML 2 baseline (2000) method. Environmental impacts from the three different MSW treatment facilities are analyzed based on 10 different impact categories in CML methods. Impact categories in CML method are abiotic depletion, acidification, eutrophication, global warming potential (GWP), ozone layer depletion, human toxicity, fresh aquatic ecotoxicity, marine aquatic ecotoxicity, terrestrial ecotoxicity, and photochemical oxidation. Characterization values of the each impact categories are analyzed; normalization of the impact category based on global value. Normalization values are taken as the world 1990 value in the LCA model and value are given in Table 1.4.

TABLE 1.4   Normalization value used in CML 2 method

Impact Categories	Unit	World, 1990
Abiotic depletion	kg Sb eq	6.32E-12
Acidification	kg SO_2 eq	3.09E-12
Eutrophication	kg PO_4^{---} eq	7.53E-12
Global warming potential (GWP100)	kg CO_2 eq	2.27E-14
Ozone layer depletion	kg CFC-11 eq	8.76E-10
Human toxicity	kg 1,4-DB eq	1.67E-14
Fresh water aquatic ecotox.	kg 1,4-DB eq	4.83E-13

TABLE 1.4   *(Continued)*

Impact Categories	Unit	World, 1990
Marine aquatic ecotoxicity	kg 1,4-DB eq	1.32E-15
Terrestrial ecotoxicity	kg 1,4-DB eq	3.79E-12
Photochemical oxidation	kg C_2H_4	5.59E-12

Source: **Pré** Consultants (2008)

1.3   RESULTS AND DISCUSSION

Comparative LCA model of pyrolysis–gasification, incineration and landfill has been developed where impact of transportation system is not considered for any of the processes.

Table 1.5 shows the characterization value of different impact categories. From the characterization table, all of the MSW treatment facility has the positive environmental impact on abiotic and ozone layer depletion categories due to the electricity generation by the processes. Landfill has the higher safety value in abiotic depletion and incineration has the higher value in ozone layer depletion category than the pyrolysis–gasification process. From the comparative study, incineration has the higher environmental impact than the Pyrolysis–gasification in the acidification, eutrophication, global warming, human toxicity, aquatic toxicity categories. However, pyrolysis–gasification has the higher potential environmental impact in terrestrial ecotoxicity and photochemical oxidation categories. Incineration has the highest GWP among the three facilities and pyrolysis–gasification has the lower GWP however, carbon emission assumed same as incineration and this was because of lower final residue production. Landfill has the highest photochemical potential among the three and incineration has the least photochemical oxidation potential. Figure 1.5 shows the characterization graph of the comparative LCA model.

TABLE 1.5   Comparative characterization model for treatment facilities

Impact category	Unit	Pyrolysis–gasification	Incineration	Landfill
Abiotic depletion	kg Sb eq	-0.04597	-0.04563	-0.09049
Acidification	kg SO_2 eq	0.24779	0.584653	0.243961
Eutrophication	kg PO_4--- eq	1.129403	1.751102	0.088294
Global warming (GWP100)	kg CO_2 eq	412.1348	424.4022	746.4556
Ozone layer depletion (ODP)	kg CFC-11 eq	-1.4E-05	-1.9E-05	-9.6E-06
Human toxicity	kg 1,4-DB eq	805.5721	1178.666	8.149164
Fresh water aquatic ecotoxicity	kg 1,4-DB eq	215.3661	323.0821	-0.25392
Marine aquatic ecotoxicity	kg 1,4-DB eq	187215.3	281106.3	835.6577
Terrestrial ecotoxicity	kg 1,4-DB eq	2.507963	0.703079	0.009382
Photochemical oxidation	kg C_2H_4	-0.00244	-0.00778	0.116526

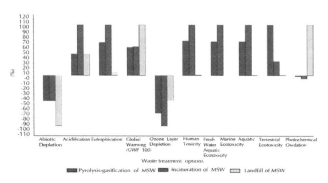

Comparing 1E3 kg 'Pyrolysis-gasification of MSW', 1E3 kg 'Incineration of MSW' and 1E3 'Landfill of MSW'; Method: CML 2Baseline 2000 V2.04/World. 1990/Characterization

FIGURE 1.5   Comparative LCA characterization graph for different waste treatment options.

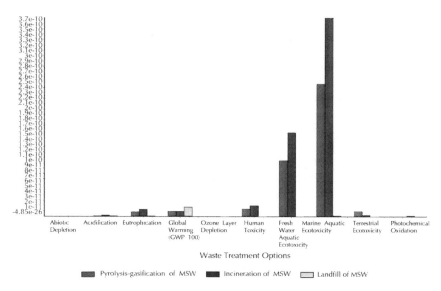

Comparing 1E3 kg 'Pyrolysis-gasification of MSW', 1E3 kg 'Incineration of MSW' and 1E3 'Landfill of MSW'; Method: CML 2baseline 2000 V2.04/World, 1990/Characterization

FIGURE 1.6   Comparative normalization graph for different MSW treatment options.

TABLE 1.6   Normalization value of the different impact categories

Impact category	Unit	Pyrolysis–gasification	Incineration	Landfill
Abiotic depletion	kg Sb eq	-2.9E-13	-2.9E-13	-5.7E-13
Acidification	kg SO_2 eq	7.66E-13	1.81E-12	7.54E-13
Eutrophication	kg PO_4^{-3} eq	8.5E-12	1.32E-11	6.65E-13
Global warming (GWP100)	kg CO_2 eq	9.36E-12	9.63E-12	1.69E-11
Ozone layer depletion (ODP)	kg CFC-11 eq	-1.3E-14	-1.7E-14	-8.4E-15

TABLE 1.6   *(Continued)*

Impact category	Unit	Pyrolysis–gasification	Incineration	Landfill
Human toxicity	kg 1,4-DB eq	1.35E-11	1.97E-11	1.36E-13
Fresh water aquatic eco-toxicity	kg 1,4-DB eq	1.04E-10	1.56E-10	-1.2E-13
Marine aquatic ecotoxicity	kg 1,4-DB eq	2.47E-10	3.71E-10	1.1E-12
Terrestrial ecotoxicity	kg 1,4-DB eq	9.51E-12	2.66E-12	3.56E-14
Photochemical oxidation	kg C_2H_4	-2.3E-14	-7.5E-14	1.12E-12

Normalization graph (Figure 1.6 and Table 1.6) shows that marine aquatic, fresh water aquatic potential, and GWP are the most significant impact categories for MSW treatment by these three facilities considering regional impact values. Normalization value shows that incineration has the higher environmental impact in marine aquatic, fresh water aquatic potential, GWP, human, and eutrophication categories than the pyrolysis–gasification processes. However, pyrolysis–gasification has the higher environmental impact in terrestrial ecotoxicity than the incineration processes. From the inventory analysis of the impact categories, vanadium, ion, selenium, nickel ion, and copper ion are the prime pollutants emitted through waste and leachate and hydrogen fluoride, benzene, carbon dioxide carbon monoxide, methane sulfur dioxide phosphate nitrogen oxide are the primary pollutants emitted to the atmosphere from the waste treatment facilities. Mercury, nickel, cadmium, hydrogen fluoride are the leading pollutants that emitted from the MSW treatment processes through air emission and cause the terrestrial ecotoxicity. Disposal of the final residue are founded as one of the most environmental impact causing phase of waste management system and vanadium, selenium, nickel copper, antimony are the leading pollutant which mainly pollutes through water and cause aquatic depletion and human toxicity. Carbon monoxide, carbon dioxide, and methane have the GWPs and photochemical oxidation, in waste management system mainly transportation of waste, processes, and disposal have the significant GWP. Pollutants though the water emissions are mainly cause eutrophication.

Global warming, acidification, and ozone layer depletion are the important impact categories considering current environmental importance. Present climate change impact acts as one of the main driving forces for sustainable decision making process.

Both incineration and sanitary landfill has the highest GWP due to CO_2 and methane emission to the atmosphere. For landfill, methane emission control of the landfill site is very difficult and costly processes. Incineration uses air for the thermal process and produce large amount of syngas during waste treatment process which is also produce large amount of CO_2. Incineration has highest acidification impact among the three due to SO_x and NO_x emission to the air. However, incineration is significantly environmental favorable to the ozone layer depletion among the three treatment options. In photochemical oxidation, landfill has highest impact among all the technologies. However, global leading impact categories (global warming or acidification) have moderately lower impact value in the normalization of LCA model. Normalization graph shows that, aquatic depletion, human toxicity occurred more from the waste treatment technologies than the other impact categories. Inventory of the model shows that, residue disposal to the landfill mainly causes aquatic depletion through ground and surface water pollution. Heavy metals pollute the environment significantly from all of the technology due to manage final residue. Landfill and incineration technologies are very old and extensively used technology. Pyrolysis–gasification is an emerging technology for municipal solid waste treatment. Therefore, comparing all these technologies through a LCA model; it is important to consider the applicability and problem solving capacity of the individual technology. Sanitary landfill found environmental favorable among the three, however, land requirement, economic, use perspective (single) and life span (around 100 years or more), landfill is not favorable in the long term perspective. That is the one of the reason of banning of landfill for different waste categories in many developed countries. On the other hand, pyrolysis–gasification is an emerging technology with high electricity production capacity from the waste.

The process is also continuous and has the option of rapid improvement in future. These factors that have been discussed before are the influential factors for the decision-making process for waste management technology selection.

1.3.1 Uncertainty and Limitations of the Results

Modern sanitary landfill with flare gas collection system for electricity generation facility has been considered for the comparison which might not be common for all countries. Sanitary landfill is more environmental friendly however; ordinary landfill has significantly high impact than the other technology. The study is done based on the process LCA analysis which is not based on waste fraction. Because assumption is made that 1 ton of waste is treated by the three different technologies and based on the emissions and energy production environmental performance of the technology is analyzed in the study. Maturity of the technology is a vital point while comparing different technologies. However, this comparative study showed the environmental burden and benefits based on the real time scale with different development level of technology. Therefore, the study did not rank any technology based on the analysis.

1.4 CONCLUSION

Different waste treatment options have different type of impacts, however, environmental soundness of the technology should be accounted in the long time perspective. Pyrolysis–gasification has found one of the emerging technologies which have lower environmental impact than the incineration process. Sanitary landfill with energy generation has the least environmental impact among the three waste treatment technologies. However, due to the socio-economic and environmental perspective landfill is not favorable waste treatment option. Disposal of final residue is one of the prime environmental concerns in thermal waste treatment processes.

KEYWORDS

- **Environmental assessment**
- **Incineration**
- **Pyrolysis–gasification**
- **Sanitary landfill**
- **Waste-to-energy**

REFERENCES

Barton, J. R. and Patel, V. S. Life cycle assessment for waste management. *Waste Manage.*, **16**(1–3), 35–50 (1996).

Bilitewski, B. and Winkler, J. Comparative evaluation of life cycle assessment models for solid waste management. *Waste Manage.*, **27**(8), 1021–1031 (2007).

Björklund, A. *Environmental systems analysis of waste management: Experiences from applications of the ORWARE model*, Ph.D. Thesis, Division of Industrial Ecology. Royal Institute of Technology, Stockholm (2000).

Björklund, A. and Finnveden, G. Life cycle assessment of a national policy proposal – The case of a Swedish waste incineration tax. *Waste Manage.*, **27**(8), 1046–1058 (2007).

Bridgwater, A. V. Catalysis in thermal biomass conversion. *Appl. Catal. A.*, **116**(1–2), 5–47 (1994).

Cherubini, F., Silvia Bargigli, S., and Sergio Ulgiati, S. Life cycle assessment (LCA) of waste management strategies: Landfilling, sorting plant and incineration. *Energy*, **34**(12), 2116–2123 (2008).

Christensen, T. H., Bhander, G., Lindvall, H., Larsen, A. W., Fruergaard, T., Anders, D., Manfredi, S., Boldrin, A., Riber, C., and Hauschild, M. Experience with the use of LCA-modelling (EASE-WASTE) in waste management. *Waste Manage. Res.*, **25**(3), 257–262 (2007).

Circeo, L. J. Plasma arc gasification of municipal solid waste, plasma applications research program, Georgia Technology Research Institute (presentation slides), Available on: http://www.energy.ca.gov/proceedings/2008-ALT-1/documents/2009-02-17_workshop/presentations/Louis_Circeo-Georgia_Tech_Research_Institute.pdf (2009).

Consonni, S., Giugliano, M., and Grosso, M. Alternative strategies for energy recovery from municipal solid waste, Part A: Mass and energy balances. *Waste Manage.*, **25**(2), 123–135 (2005).

DEFRA (Department for Environment, Food, and Rural Affairs). *Review of environmental and health effects of waste management: Municipal solid waste and similar wastes*, accomplished by Enviros Consulting Ltd. and University of Birmingham with Risk and Policy Analysts Ltd., Open University and Maggie Thurgood (2004).

DEFRA (Department for Environment, Food and Rural Affairs). Incineration of municipal solid waste, waste management technology brief, the new technologies work stream of the Defra Waste Implementation Programme (2007).

Diaz, R. and Warith, M. Life-cycle assessment of municipal solid wastes: Development of the WASTED model. *Waste Manage.*, **26**(8), 886–901 (2006).

Ekvall, T., Assefa, G., Björklund, A., Eriksson, O., and Finnveden, G. What life-cycle assessment does and does not do in assessments of waste management. *Waste Manage.*, **27**(8), 989–996 (2007).

Ekvall T. and Finnveden G. The application of life cycle assessment to integrated solid waste management, Part 2: Perspectives on energy and material recovery from paper. *Trans. IChemE.*, **78**(4), 288–294 (2000).

Farquhar, C. J. and Rovers, F. A. Gas production during refuse decomposition. *Water Air Soil Pollut.*, **2**(4), 483–495 (1973).

FCM (Federation of Canadian Municipalities). Solid waste as a resource, review of waste technologies, 111. Available on: http://www.sustainablecommunities.ca/files/Capacity_Building__Waste/SW_Guide_Technology.pdf (2004).

Feo, G. D., Belgiorno, V., Rocca, C. D., and Napoli R. M. A. Energy from gasification of solid wastes. *Waste Manage.*, **23**(1), 1–15 (2003).

Finnveden, G., Johansson, J., Lind, P. and Moberg, A. Life cycle assessments of energy from solid waste, project report of "Future oriented life cycle assessments of energy from solid waste" project, FMS report (2000).

Finnveden, G. and Moberg, A. Environmental system analysis tools: An overview. *J. Clean. Produt.*, **13**(12), 1165–1173 (2004).

Gheewala, S. H. and Liamsanguan C. LCA: A decision support tool for environmental assessment of MSW management systems. *J. Environ. Manage.*, **87**(1), 132–138 (2008).

Grieco, E. and Poggio, A. Simulation of the influence of flue gas cleaning system on the energetic efficiency of a waste-to-energy plant. *Appl. Energ.*, **86**(9), 1517–1523 (2009).

Hallenbeck, W. H. Health impact of a proposed waste to energy facility in Illinois. *Bull. Environ. Contam. Toxicol.*, **54**(3), 342–348 (1995).

Halton EFW Business Case. The regional municipality of Halton, Step 1B: EFW Technology Overview (2007).

Khoo, H. H. Life cycle impact assessment of various waste conversion technologies. *Waste Manage.*, **29**(6), 1892–1900 (2009).

Liamsanguan, C. and Gheewala, S. Environmental assessment of energy production from municipal solid waste incineration. *Int. J. LCA.*, **12**(7), 529–536 (2007).

Ludwing, C., Hellweg, S., and Stucki, S. Municipal solid waste management; strategies and technologies for sustainable solutions, Springer (2002).

Malkow, T. Novel and innovative pyrolysis and gasification technologies for energy efficient and environmentally sound MSW disposal. *Waste Manage.*, **24**(1), 53–79 (2004).

Matsuto, T. Life cycle assessment of municipal solid waste management-Cost, energy consumption and CO_2 emission, Proceedings of International Symposium and Workshop on Environmental Pollution Control and Waste Management 7–10 January 2002, Tunis (EPCOWM'2002), pp. 243–248 (2002).

NSCA. Comparison of emissions from waste management options. National Society for Clean Air and Environmental Protection (2002).

Parizek, T., Bebar, L., and Stehlik, P. Persistent pollutants emission abatement in waste-to-energy systems. *Clean Tech. Environ. Policy*, **10**(2), 147–153 (2008).

Parker, A. Behaviour of Wastes in Landfill-Leachate, Chapter 7, Behaviour of Wastes in Landfill-Methane Generation, Chapter 8, J. R. Holmes (Ed.). Practical Waste Management, John Wiley and Sons, Chister, England (1983).

Pennington, D. W. and Koneczny, K. Life cycle thinking in waste management: Summary of European Commission's Malta 2005 workshop and pilot studies. *Waste Manage.*, **27**(8), 592–597 (2007).

Pohland, F. G. Critical review and summary of leachate and gas production from landfills. EPA/600/S2-86/073, *U.S. EPA*, Hazardous Waste Engineering Research Laboratory, Cincinnati, OH (1987).

Pohland, F. G. *Fundamental principles and management strategies for landfill codisposal practices.* Proceedings Sardinia 91, 3rd International Landfill Symposium, Vol. 2, pp. 1445–1460, Grafiche Galeati, Imola, Italy (1991).

Pré Consultants. SimaPro 7, Amersfoort, The Netherlands (2008).

Psomopoulos, C. S., Bourka, A., and Themelis, N. J. Waste-to-energy: A review of the status and benefits in USA. *Waste Manage.*, **29**(5), 1718–1724 (2009).

Stehlik, P. Contribution to advances in waste-to-energy technologies. *J. Clean. Product.*, **17**(10), 919–931 (2009).

Tehrani, S. M., Karbassi, A. R., Ghoddosi, J., Monavvari, S. M., and Mirbagheri, S. A. Prediction of energy consumption and urban air pollution reduction in e-shopping adoption. *Int. J. Food, Agri. Environ.*, **7**(3–4), 898–903 (2009).

2 Solid Waste Management in a Mexican University Using a Community-based Social Marketing Approach

Carolina Armijo-de Vega*, Sara Ojeda-Benítez, Quetzalli Aguilar-Virge, and Paul A. Taboada-González

CONTENTS

*   Address correspondence to this author at the Facultad de Ingeniería Ensenada, Universidad Autónoma de Baja California, Km 103 Carretera Tijuana-Ensenada, Ensenada, Baja California, C.P. 22870, México; Tel/Fax: 52(646) 174-4333; E-mail: carmijo@uabc.edu.mx

2.1 INTRODUCTION

Waste separation and recycling programs in higher education institutions requires an approach that reach people in different ways. Social marketing approach has proved to be effective in helping reach the desired change for very different initiatives. This chapter presents a 16 month experience of a paper and cardboard separation program at the Ensenada Campus of the Autonomous University of Baja California (UABC). Although, the support from the University authorities is important, through different experiences it was found that in UABC the programs that work better are the ones that do not depend on the work of personnel but on the participation of students and academic staff. To gain this participation, the strategies used in social marketing were used. To date through UABC's paper and cardboard program, the institution has diverted more than 6 tons of this type of waste from the main waste stream. Based on the evaluation of the program and on the increasing community response, it can be said that the social marketing strategies used in this program were successful.

In the 21st century, higher education institutions have to face a series of challenges such as the promotion and implementation of sustainable practices through the participation of faculty, students, and staff, which should be compromised in building a better future for the generations to come. Diverse research have shown that the role that universities and their faculty play when promoting sustainable practices is key and influences the success of other sustainability programs in society (Ferrer-Balas et al., 2009; Juárez-Najera et al., 2006; Lehmann et al., 2009; Velázquez et al., 2006; Zilahy and Huisingh, 2009).

In addition, several universities worldwide have incorporated the sustainability approach to their courses and academic programs to form professionals sensible to environmental protection (Crofton, 2000; Ferrer-Balas et al., 2009; Lidgren et al., 2006; Ramos, 2007). Education and formation of new professionals must include the sustainable approach as to acquire the necessary skills to face diverse environmental problems. In this sense, universities should put into practice strategies for sustainable development which must be immersed in their academic programs, research, outreach, and facilities operation. One of the many environmental problems that must be addressed is the one related to the increasing amounts of solid waste.

Internationally, municipalities are challenged every day with the complexity of solid waste management; the increasing generation of waste, the limited resources available for its management and the lack of responsibility from waste generators worsen the problem. This implies that problems generated as a consequence of the improper management of municipal solid waste (MSW) are complex because waste is generated in diverse sectors such as commercial (stores), education (schools), health (hospitals), recreation (parks), and touristic (hotels), among others. These establishments are heterogeneously distributed in the cities and have different performance contexts, as well. This diversity of waste generators makes very difficult to implement effective and efficient waste management initiatives. To face this complexity in the management of MSW some countries have put into practice sector-tailored solid waste strategies. In this way, waste generators of the same section or sector, for example the hotel section, get organized and create common plans for waste management that includes common practices for the segregation by waste type, for temporal storage, transport, and treatment. Through, this organization the responsibility of waste management is shared among the same section generators.

To achieve sectional waste management plans, first it is necessary to know the characteristics of waste that each section generates and the approximate amounts. Also, it is necessary to implement waste management pilot programs to detect and correct possible failures and to add new practices that could improve the program in each section. In this sense, recent research carried out in different parts of the world show that colleges and universities are not aside from the problems related with waste generation. For this reason some institutions have involved in waste management programs with the objective to recover the recyclable materials (Armijo-de Vega et al., 2003; Kelly et al., 2006; Masson et al., 2004), in the implementation of zero waste programs in university campus (Masson et al., 2003), and on recovering of paper (Amuteya et al., 2009). Moreover, some educational institutions have also engaged in the promotion of a new conception of man and nature through a change in attitudes, culture, and consciousness; in this sense a research was carried out to know the attitudes and behavior towards recycling in a university campus (Kelly et al., 2006).

It is also important to mention that in an education institutions waste composition is different from household waste; Table 2.1 shows the differences in composition by weight of these two generation sources.

TABLE 2.1   Waste Composition in Education Institution and Household

Type of Waste	Education Institutions (Percentage of Total Waste)	Households (Percentage of Total Waste)
Paper and cardboard	20–50 % [1]	11–20% [11]
Organics	20–48% [111]	22–55% [111]

Observing these differences, it is imperative to know the quantities and characteristics of waste generated in each section before proposing sectional waste management plans; this is also valid for the school section.

Different experiences have shown that the logistics and technology alone are not enough for a recycling program to be successful because the human factor plays a key role. Thus, an important component of any recycling program is the communication and information campaign that seeks to reach the people intended to participate. In the Autonomous University of Baja California (UABC), diverse recycling initiatives have taken place since 1998[*] but these have been focused on the logistics of recyclables separation. Solid waste characterization studies have also taken place at UABC in order to propose an integrated solid waste management program. Information campaigns for these initiatives had the objective to make clear how to use the different recycling bins, where these new bins were located and how separation should be made.

Information was the same for all audiences in the organization and was delivered through an Internet site, flyers, radio and TV spots, and conferences. One common characteristic of previous recycling initiatives was that all were announced and promoted by people in top-management positions of the university. Other common thing among these programs was that they all depended on the participation of maintenance

---

[*]   Arroyo, V., personal communication, 2006.

staff. One last element shared in previous programs was their financial dependence for the logistics and publicity of the programs. Despite the effort made in the information campaigns and in the logistics, those recycling programs in UABC did not bring effective results. Waste was not properly separated, bins were not used the way they were supposed to, people were not keen to participate in the programs, and finally, it was very difficult to sell the recyclables because of these problems.

In view of the results of the previous recycling efforts, a different approach was used to promote and impel a new initiative to separate and recycle paper and cardboard at the Campus Ensenada of UABC. The study reported here has the objective to implement and evaluate the performance of a paper and cardboard segregation program in one of the campuses of the Autonomous University of Baja California. This program was planned and undertaken through a social marketing strategy (Kotler and Lee, 2007; McKenzie-Mohr and Smith, 1999). Based on the experiences of the initiative reported here, the program is expected to expand to all the campuses of UABC and later to other colleges and universities of Baja California. Finally, these experiences could set ground for a school sector oriented waste management strategy.

2.2   SOCIAL MARKETING APPROACH

Advocacy messages commonly face the challenge of trying to change behavior by forcing consumers to confront some disconcerting reasons for the need to abandon the *status quo*. Ads or other message forms appealing for increased recycling are no exception (Lord and Putrevu, 1998). According to Hopper and Nielsen (Hopper and Nielsen, 1991), recycling is an altruistic behavior; and De Young (DeYoung, 1990) mentions that efforts to promote waste reduction and recycling behavior should focus on non-monetary motives. The question then is how to appeal to non-monetary motives to make people participate in waste separation recycling programs? Social marketing offers an alternative approach to the typical information channels for recycling programs such as, flyers, TV spots, posters, stickers, and so on. Community base social marketing is based upon research in the social sciences that demonstrates that behavior change in most effectively achieved through initiatives delivered at the community level, which focus on removing barriers to an activity while simultaneously enhancing the activities benefits.

Social marketing arose as a discipline in the 1970s, when Philip Kotler and Gerald Zaltman (Kotler and Zaltman, 1971) realized that the same marketing principles that were being used to sell products to consumers could be used to "sell" ideas, attitudes, and behaviors. Like commercial marketing, the primary focus is on the consumer on learning what people want and need rather than trying to persuade them to buy what we happen to be producing. The application of marketing principles and techniques to promote a social cause, idea or behavior has been effectively used in many recycling programs (Birgonia, 2007; Cole, 2007; MacLennan and McConnell, 2007; Shrum et al., 1994; Tabanico and Schultz, 2007). Social marketing approach has been found to significantly contribute to the attainment of specific program objectives and goals. Implementing it, however, involves a decision by management to undertake a focused and purposive activity requiring the kind of support that is anchored on the belief that this approach in fact, can make a difference (Birgonia, 2007).

Social marketing has emerged as an alternative to promote environmentally friendly practices such as recycling (Ball, 2008; Tabanico and Schultz, 2007; Werder, 2005). It is a unique approach because it offers a framework for the people in need to promote behavioral changes in diverse establishments. Community-based social marketing also uses tools that have been identified as being particularly effective in fostering change. Although, each of these tools on its own is capable of promoting sustainable behavior, the tools can often be particularly effective when used together. Key community-based social marketing tools include:

- *Prompts:* Numerous behaviors that support sustainability are susceptible to forgetting. Prompts can be very effective in reminding to perform certain activities remind people to engage in sustainable activities (e.g., a vehicle window sticker indicating that the driver does not idle);

- *Commitment:* According to McKenzie-Mohr and Smith in a wide variety of settings people who have initially agreed to a small request, have subsequently been found to be far more likely to agree to a larger request. These authors recommend having people commit or pledge to engage in sustainable activities through, for example, signing a pledge card to avoid unnecessary idling.

- *Communication:* Programs that intend to foster sustainable behavior should include a communication component. In this program short e-mail messages were used with relevant information about the progress of the recycling program such as quantities of cardboard on paper separated, punctual instructions on how to separate, and the e-mail and phone number of the people in charge. Special attention was paid to the recommendations of McKenzie-Mohr and Smith (1999) in relation to the usage of captivating information, credible sources, avoiding the use of threatening messages, use of massages easy to remember, among other.

- *Removing external barriers:* The behavior change strategies used in social marketing can have a significant influence upon the adoption and maintenance of behavior.

- *Norms:* Develop community norms that a particular behavior is the right thing to do; and,

- *Incentives:* These are used to reward people for taking positive actions, such as returning beverage containers, rather than fining them for engaging in negative actions. Incentives can be powerful levers to motivate behavior.

Social marketing starts with the selection of a "target behavior" and later uses a four stage process to encourage the desired change towards a sustainable behavior. These four stages are (McKenzie-Mohr and Smith, 1999):

1. **Identifying Barriers to a Particular Behavior**
   Research indicates that each form of sustainable behavior has its own group of barriers (Allen, 1999; Bowers, 1997; Ching and Gogan, 1992; Clugston and Calder, 1999; Creighton, 1998; Dhale and Neumayer, 2001; Hamburg and Ask, 1992; Schriberg, 2002). To promote activities that support sustainability, barriers to these activities must first be identified. Community-based social marketing therefore begin by conducting the research that will help to identify

these barriers. It is not unusual to uncover multiple barriers that are quite specific to the activity being promoted. Once the barriers have been identified, the next step is to develop a program that addresses each of them. Personal contact, the removal of barriers and the use of proven tools of change are emphasized in the program.

2. **Developing and Piloting a Program to Overcome these Barriers**
   To ensure that the program will be successful, it should be piloted in a small segment of the community and refined until it is effective. The program is then implemented throughout the community, and procedures are put in place to continually monitor its effectiveness.

3. **Implementing the Program across a Community**
   The steps that make up community-based social marketing are simple but effective. When barriers are identified and appropriate programs are designed to address these barriers, the frequent result is that individuals and organizations adopt more sustainable activities which is the cornerstone of healthier and more sustainable communities.

   Social marketing is based in social sciences research and particularly in psychology that has identified a variety of effective tools to promote behavior change. These tools are more effective when used combined.

4. **Evaluating the Effectiveness of the Program**
   In order to know the degree of success of the strategies for change, it is necessary to evaluate the implementation of the program by obtaining information on baseline involvement in the activity prior the implementation and at several points afterward.

2.3 METHOD

The paper and cardboard recycling pilot program reported here took place at the Campus Ensenada of the Autonomous University of Baja California, Mexico. The evaluation of the program was made during 16 months from January 2008 to May, 2009, July is not considered because is the summer vacation and no waste is generated during that time.

The steps followed to implement the mentioned program were the ones proposed by McKenzie-Mohr and Smith (1999), these are the following:

2.3.1 Identifying Target Behavior

The first step to implement the program starts with two questions. The first question that was addressed in this step was *what behavior should be promoted?* To decide which behavior to promote at UABC it was necessary to answer the question, what is the potential of an action to bring about the desired change? To answer this, a detailed analysis was made about the desired change. This analysis was made based on the previous waste management experiences at UABC and on the present day institutional context.

A second question that had to be answered in this stage was, *who should the program address or target?* To answer this question it was made a review of the results of a previously applied questionnaire aimed to detect the attitudes towards reducing, reusing, and recycling waste. This questionnaire was applied to a sample group of the university community that included students, administrative staff, faculty, and custodians.

2.3.2 Identifying Barriers to a Particular Behavior

To detect the barriers to separate cardboard and used paper three steps were followed:

1. *Literature review:* Academic books and articles were reviewed in order to detect to most typical barriers encountered in other places when new waste management programs were implemented .
2. *Observation and interviews:* Qualitative information campaign a survey was made which included was obtained through observation of the way people seven questions with five Likert scale values each, where 1 working or studying at UABC generate and disposes corresponded to total disagreement, 3 to a neutral position paper and cardboard. Twenty persons were interviewed (five secretaries, five students, five custodians and five professors). The objective of the interview was to recognize why they were handling their waste that way and if they had any knowledge about the implications of their behavior.
3. *Survey:* A survey was constructed, validated and applied to 30 people randomly chosen. The objective of the survey was to identify the attitudes towards waste management and the disposition to participate in a paper and cardboard separation program. It was also asked how they would prefer to receive information about a waste management program.

2.3.3 Use of Tools for Behavior Change

1. *Commitment:* For this program, diverse types of commitment were sought: written, verbal, public, group, actively involving a person.
2. Different prompts were used such as signs in the offices, signs near the recycling bins, and short written explanations about the characteristics of waste to be recycled.
3. *Communication:* Diverse communication strategies were used such as conferences, flyers, stickers, e-mail reminders and information messages. The first two were delivered through the work of social service students, the latter through the University mass e-mail service.
4. *Incentives:* Even though incentives have shown to have an important impact in a variety of programs to recycling, in this program incentives were not used because the program lacked financial support.
5. Based on the barriers detected in the previous stage (identifying barriers to a particular behavior) two main strategies were used:
   a. Location of recycling bins in convenient places near the paper generation points.
   b. Twice a week collection of the materials separated in the recycle bins. The collection was made by social service students.

2.3.4   Design and Evaluation

Once the barriers were identified and prioritized, the change tools that matched the barriers were selected. Feedback from the participants was obtained and latter the pilot program was launched in two faculties. The pilot program was functioning during 16 months, some failures were corrected, and then it was expanded to all the faculties of the campus.

To evaluate the general progress of the separation of paper and cardboard, the monthly quantities of these materials were recorded.

To evaluate the efficiency of the communication and information campaign a survey was made which included seven questions with five Likert scale values each, where 1 corresponded to total disagreement, 3 to a neutral position and 5 to a total agreement position. This survey was applied to 40 people in three different times: 3, 12, and 16 months after the implementation of the program. The questions included in the survey were the following:

1. The waste that I generate is my responsibility
2. UABC is an institution that manage its solid waste properly
3. Paper separation program at UABC promotes a culture of environmental responsibility
4. I am willing to actively participate in the paper and cardboard separation program of UABC
5. Is easy and convenient to separate paper
6. I know the location of the paper bins
7. I am informed about the progress and changes of the paper and cardboard program at UABC.

2.4   RESULTS

The results are presented in the same order as the steps presented in the methodology section.

2.4.1   Identifying Target Behavior

All previous waste programs at UABC presented an inadequate separation of waste; this problem was also identified by other waste management coordinators of other universities (Armijo-de Vega, 2003; Florida State University, 2007; Keniry, 1995). Thus, the target behavior identified was "an adequate diversion of paper and cardboard". By adequate we refer to the separation of materials that do not include contaminants or other types of waste but the ones indicated by the program.

A second target behavior was the "correct disposition of paper and cardboard in the containers destined to deposit those materials". This was decided since in previous experiences of recycling in UABC one of the main problems was that although the generators of residues knew well the type residues that had to be deposited in recycling containers the disposal was incorrectly performed.

The objective population to which the campaign would go was academic and administrative personnel, the participation of students occurred indirectly. This was decided because the former generate more paper and cardboard in the campus and are the groups that can be monitored for longer periods. While, students remain less time in the university facilities and leave after three or four years, so is more difficult to follow their recycling behavior.

2.4.2   Identifying Barriers to a Particular Behavior

The information found in literature agreed with the findings of this study as far as the barriers to make the desired change, in the discussion section are mentioned these. Table 2.2 shows a simple matrix that presents the perceived benefits and barriers, as well as the behaviors that compete with the target behavior detected during the observations and interviews.

Tables 2.2 and 2.3 show the perceived benefits and barriers more frequently mentioned during the interviews. These results made evident that the strategies should be oriented to facilitate the process for material separation and disposition. To achieve these goals, two different types of temporal disposal sites were placed to separate paper and cardboard: (1) primary sites and (2) secondary sites. The former were Gaylord boxes (47" × 36" × 50") which were intended for the temporarily store of considerable quantities of paper and cardboard. These boxes were located in sites protected from rain and wind but at the same time that were accessible (Figure 2.1) for the deposition of material and for the collection.

The secondary deposit sites were located near the paper and cardboard generation sites, mainly inside offices or in corridors. For example for a group of cubicles a median size box was located in the corridor (Figure 2.2). If the professors wanted and space was available, a small box was placed as well in his or her office (Figure 2.3), so that they did not have to move to deposit any material. This was only made in the cases in which the box for paper did not represent a problem of space in the office.

TABLE 2.2   Perceived benefits and barriers and competing behaviors (for target behavior 1)

	Target Behavior Correct Separation of Materials	Competing Behavior 1 Easy to Dispose Al the Materials Mixed
Perceived Benefits	Helps the environment	No need to differentiate types of waste
Perceived Barriers	Lack of time to separate waste types	Bad for the Environment Costly disposition of waste in landfill

TABLE 2.3   Perceived benefits and barriers and competing behaviors (for target behavior 2)

	Target Behavior Correct Deposition of Materials in Recycling Bins	Competing Behavior 1 Everything is Disposed in the Same Bin	Competing Behavior 2 Throwing the Waste from its Place
Perceived Benefits	Helps the environment Good image	No need to move to the recycling bin	Saves time
	Exemplary behavior Lack of recycling bins		
Perceived Barriers	Lack of space for recycling bins	Bad for the environment	Bad image

The collection of paper in the secondary disposal sites was carried out by social service students. Students picked-up the materials and deposit them in the primary disposal sites. This way the perceived barriers mentioned by faculty and staff would be overcome. Each time the primary deposits were full, a recycling company was called to collect the materials.

FIGURE 2.1   Primary disposal sites for paper and cardboard.

FIGURE 2.2   Secondary disposal sites in corridors.

2.4.3 Use of Tools for Behavior Change

In this study the following tools were used:

- Verbal commitment was emphasized in offices, group commitment was pursued, people was actively involved, coercion was never used, people was helped to see themselves as environmentally responsible. This was made in 26 administrative offices, in the four faculties of the campus and in the two institutes.
- Visual prompts were used in corridors, primary disposal sites, and *via* e-mail. The prompts were mainly used to remind the types of material to be separated and the location of the temporal disposal sites.
- The communication system was through e-mail, this media was chosen because it can be massively delivered to the whole campus. The information delivered was focused on the quantities of materials diverted from the main waste stream, a short explanation of the program and a thank you note to let the people know that the success of the program was because of the community participation and commitment.
- The elimination of barriers consisted on the convenient location of the primary and secondary disposal sites and in the collection made by students. This way, the participants would only have to correctly separate cardboard and paper and students would collect the materials from the generation site.

FIGURE 2.3 Secondary disposal sites in cubicles.

2.4.4   Evaluation

Every 2 weeks social service students reported the conditions of use of primary and secondary disposal sites and the number of complaints received. This way any inconvenience or misuse of boxes could be corrected, and signs replaced.

A hard indicator of the progress of the paper and cardboard program was the monthly amount (kg) of materials separated. For doing this, the quantities of the materials diverted were recorded; (Figure 2.4) shows the monthly quantities (kg) of paper and cardboard in a 16 month period. The total amount for the 16 month period is 6,008 kg of paper and cardboard.

Figure 2.4 shows the quantities of paper and cardboard have fluctuated during the evaluated period. The first 2 months present the lowest amounts of materials because the program then was present only in two faculties. During the first semester, the highest amount was reach in April that coincided with the expansion of the program to all the campus. Also, in April one of the campus faculties made an aggressive campaign involving a group of students inviting the university community to clean their offices and get rid of old notes and exams. The effect of this campaign lasted until May and decreased in June. The month of July reports no results because is the summer vacation period and no activity takes place at campus. The first year (from January to December, 2008) had a monthly average of 295 kg; the next five months (January to May, 2009) had an average of 552 kg. This difference shows an average increase of 87% the amount of the first year.

It is to be noted that the reported quantities were informed by the recycling company since we did not have the equipment to weigh the materials before they were collected by the company. Because the separation and collection of paper and cardboard were made in a single container, the data of both materials are reported together.

The survey to evaluate the efficiency of the communication and information campaign of the program showed progress. In Figure 2.5, it can be observed that the seven indicators improved with time, this means that a positive change in the perception of the program took place.

When the survey was applied for the first time (blue line) the values were low, principally in relation to the willingness to participate in the program (question 4). In general, it can be mentioned that the perception of participants in relation to their responsibility as waste generators changed positively (question 1), although a decrease of one unit is present for the last evaluation (green line). A positive trend was shown in the perception that UABC manage its solid waste properly (question 2). The perception that UABC promotes a culture of environmental responsibility (question 3) was at its higher value since the second time the survey was applied. The perception that is easy and convenient to separate paper (question 5) also improved, this indicates that people is realizing that this activity does not take much time and can be easily done. The knowledge of the location of the recycling bins (question 6) is a good indicator that the signs and prompts are working well. Finally, the survey showed that the people is informed about the progress of the paper and cardboard program (question 7), but special care must be paid here because the survey was applied just 1 day after the last information e-mail was sent. In general, Figure 2.5 shows a positive trend in most of the indicators, nevertheless more attention must be paid to the awareness campaign in order to improve the perception of waste generators as responsible participants of the waste problem.

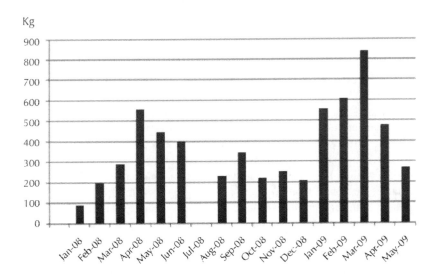

FIGURE 2.4   Quantities (kg) of paper and cardboard generated in 16 month period.

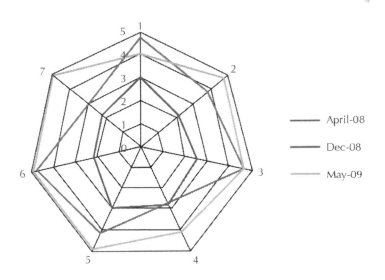

FIGURE 2.5   Results of the survey in three different times.

2.5 DISCUSSION

Recycling programs contributes to institutional solid waste reduction objectives; leading community by practicing ethical and responsible waste management. Thus, also special attention should be focused on the strategies used to involve and influence more people to participate in the program. At UABC, the paper and cardboard recycling program is not mandatory, this is the reason why not all the staff, faculty, and students are participating. Although, the authors are positive that more people will get involved in the program if social marketing strategies continue to be used.

Some social contexts may actively discourage the widespread adoption of recycling activities. Individual concern for the environment and individual resources such as education may not easily overcome contextual barriers to action. However, if the context is changed to facilitate the adoption of new behavior, the probability of individual action should increase because the effort required on the part of any single individual decreases. In the case reported here, the context change was the convenient location of recycling bins and the materials collection made by students. Under these circumstances, recycling would require relatively little effort, and as a consequence, participation is further promoted more. This finding agrees with Taylor and Todd (1995) who found that a similar concept self-efficacy (the perceived ability to carry out the behavior), leads to perceived behavioral control and from there to a positive intention to recycle. This is also in accordance with Derksen and Gartrell (1993) in that the most important determinant of recycling behavior is access to a structured program that makes recycling easy and convenient. Amutenya, Shackleton, and Whittington-Jones (2009) also discuss the importance of removing the distance barriers by increasing the number of recycling bins which leads to a potential increase in recycling. Thøgersen (1997) supports the above, demonstrating the usefulness of social marketing approach for the promotion of recycling through the design of reverse distribution channels for recyclables.

Furthermore, a systematic, well-advertised program could create a new community norm favoring recycling. In accordance to this, one way of encouraging a long-term recycling behavior is through information and dissemination techniques (Amuteya and Shackleton, 2009). Community based social marketing (CBSM) applied to a social cause such as recycling offers a good approach for dissemination and the delivering of information. In this sense, periodic prompts, information, and follow-up surveys should be an ongoing part of the program. In the case of the paper and cardboard recycling program of UABC, this continuous approach has taken place and will continue to be. Nevertheless, more attention should be paid to the time gaps where no students are present. At present, no strategy is in place for the weeks when student finish their social service program and a new term starts. This could help to explain the reduction in paper and cardboard collected by the program during the inter-semester periods.

2.6 CONCLUSION

Clearly, recycling is an essential element of any long-term solution to the problem of waste, and it becomes critical concern for recycling advocates that how to motivate full participation in recycling programs. The CBSM as a framework to foster recycling is suited for university settings. A variety of CBSM strategies were employed at the Autonomous University of Baja California to address the issue of proper waste management.

Based on the results presented here, it can be concluded that the paper and cardboard separation program of UABC is progressing. For the case reported here, the social marketing tools proved to be effective to influence public behavior and this could be because it focused on the target audience's point of view, this made account of the emotional or physical barriers that may have prevented people from changing their behavior and not on coercion neither on fear campaigns that only have short time effects.

Although, CBSM approach has been applied to different environmental programs internationally, the interesting issue found in this study is that this is not the first attempt to implement a waste management program at UABC. The difference now, compared to the previous waste management initiatives in this institution, is that the latter were led by top management authorities using the typical command chain way to impose new practices, in this case, new ways to deal with waste. The program reported in this paper did not depend on support from authorities nor from custodians. In this sense a bottom-up program was being promoted using a completely new approach.

To facilitate the initial steps towards change is positive because this makes people to easily "hook" on the proposed activities. Nevertheless, it is imperative to also search for strategies that seek a community's deeper responsibility in waste management, not because it is easy to hook but because it is right to do it.

The paper and cardboard program of UABC is only the start of an integrated solid waste management program. Before including more categories of waste into the program (e.g., plastics or metals) an adjustment must be made to the follow-up of the quantities of material generated. A paper and cardboard separate record should be made to know the precise quantities of each material and have better indicators of the advancement of the program.

ACKNOWLEDGMENT

The authors would like to express their gratitude to all the social service students who participated in the paper and cardboard separation program during the reported period. The authors would also like to thank the staff, students, and faculty who responded the surveys and interviews during the evaluation stage of this program.

KEYWORDS

- **Social marketing**
- **Solid waste**
- **Recycling**

REFERENCES

Allen, A. S. "*Institutional environmental change at Tulane University,*" Tulane University (1999).

Amuteya, N., Shackleton, C. M., and Whittington-Jones, K. "Paper recycling patterns and potential interventions in the education sector: A case study of paper streams at Rhodes University, South Africa." *Resour. Conserv. Recycl.*, **53**, 237–242 (2009).

Armijo-de Vega, C. Ojeda-Benítez, S., and Ramírez-Barreto, M. E. "Mexican educational institutions and waste management programmes: A University case study." *Resour. Conserv. Recycl.*, **39**, 283–296 (2003).

Armijo-de Vega, C., Ojeda-Benítez, S., and Ramírez-Barreto, M. E. "Solid waste characterization and recycling potential for a university campus." *Waste Manage.*, **28**, S21–26 (2008).

Ball, S. "Social marketing as a means to influence student behavior towards energy conservation," M. Sc., University of Mary Washington (2008).

Birgonia, C. *Social marketing on solid waste management: The Jagna Experience.* EcoGov2, Philipines (2007).

Bowers, C. A. *The culture of denial: Why the environmental movement needs a strategy for reforming universities and public schools*, State University of New York Press, Albany (1997).

Buenrostro-Delgado, O. *Los residuos sólidos municipales: Perspectivas desde la investigación multidisciplinaria.* Universidad Michoacana de San Nicolás Hidalgo, Morelia, Mich. Mexico (2001).

Ching, R. and Gogan, R. "Campus Recycling: Everyone Plays a Part," *The Campus and Environmental Responsibility.* Jossey-Bass, San Francisco, pp. 113–125 (1992).

Clugston, R. and Calder, W. "Critical Dimensions of Sustainability in Higher Education," *Sustainability and University Life.* Peter Lang, New York, pp. 31–46 (1999).

Creighton, S. H. *Greening the ivory tower: Improving the environmental track record of universities, colleges and other institutions.* MIT Press, Cambridge, MA (1998).

Cole, E. J. "A community-based social marketing campaign to green the offices at Pacific University: Recycling, paper reduction and environmentally preferable purchasing." Antioch University (2007).

Cortinas-De Nava, C. *Hacia un Mexico sin basura*, Ciudad de Mexico: Partido Verde Ecologista de Mexico (2001).

Crofton, F. S. "Educating for sustainability: Opportunities in undergraduate engineering." *J. Cleaner Prod.*, **8**, 397–405 (2000).

Derksen, L. and Gartrell, J. "The social context of recycling." *Am. Sociol. Rev.*, **58**, 434–442 (1993).

DeYoung, R. "Recycling as appropriate behavior: A review of survey data from selected recycling education programs in Michigan." *Resour. Conserv. Recycl.*, **3**, 253–266 (1990).

Dhale, M. and Neumayer, E. "Overcoming barriers to campus greening." *Int. J. Sust. Higher Educ.*, **2**, 139–160 (2001).

Espinosa, R. M., Turpin, S., Polanco, G., De la Torre, A., Delfin, I., and Raygoza, I. "Integral urban solid waste management program in a Mexican university." *Waste Manage.*, **28**, S27–S32 (2008).

Felder, M. A. Petrell, R. J., and Duff, S. J. "A solid waste audit and directions for waste reduction at the University of British Columbia, Canada." *Waste Manage. Res.*, **19**, 354–365 (2001).

Ferrer-Balas, D., Buckland, H., and Mingo, M. "Explorations on the University's role in society for sustainable development through a systems transition approach". Case-study of the technical University of Catalonia (UPC)." *J. Cleaner Prod.*, **17**, 1075–1085 (2009).

Florida State University, *Recycling at Florida State University*, Miami: Florida State University, (2007).

Hamburg, S. P. and Ask, S. I. "The environmental ombudsman at the University of Kansas," *The campus and environmental responsibility.* Jossey-Bass, San Francisco, pp. 55–63 (1992).

Hopper, J. R. and Nielsen, J. M. "Recycling as altruistic behavior: Normative and behavioral strategies to expand participation in a community recycling program." *Environ. Behav.*, **23**, 195–220 (1991).

Juárez-Najera, M., Dieleman, H., and Turpin-Marion, S. "Sustainability in Mexican Higher Education: Towards a new academic and professional culture." *J. Cleaner Prod.*, **14**, 1028–1038 (2006).

Kelly, T. C., Masson, I. G., Leiss, M. W., and Ganesh, S. "University community responses to on campus resource recycling." *Resour. Conserv. Recycl.*, 47, 42–55 (2006).

Keniry, J. *Ecodemia: Campus environmental stewardship at the turn of the 21st Century.* National Wildlife Federation, Washington, DC. (1995).

Kotler, P. and Lee, N. *Social Marketing: Influencing Behaviors for Good.* Sage, Thousand Oaks CA (2007).

Kotler, P. and Zaltman, G. "Social Marketing: An Approach to Planned Social Change." *J. Market.*, 35, 3–12 (1971).

Lehmann, M., Christensen, P., Thrane, M., and Herreborg Jørgensen, T. "University engagement and regional sustainability initiatives: Some Danish experiences." *University Engagement and Regional Sustainability Initiatives: Some Danish Experiences*, 17, 1067–1074 (2009).

Lidgren, A., Rodhe, H., and Huising, D. "A systemic approach to incorporate sustainability into university courses and curricula." *J. Cleaner Prod.*, 14, 797–809 (2006).

Lord, K. R. and Putrevu, S. "Acceptance of recycling appeals: The moderating role of perceived consumer effectiveness." *J. Market. Manage.*, 14, 581–590 (1998).

MacLennan, B. L. and McConnell, R. L. "Use what you have: Strategies for developing a hybrid marketing approach," *Proceedings of the International Conference on Waste Technology and Management*, Widener University, Philadelphia, PA, pp. 716–727 (2007).

Maldonado, L. "Reducción y reciclaje de residuos sólidos urbanos en centros de educación superior: estudio de caso." *Revista Ingeniería*, 10, 59–68 (2006).

Masson, I. G. Brooking, A. K. Oberender, A., Harford, J. M., and Horsley, P. G. "Implementation of a zero waste program at a university campus." *Resour. Conserv. Recycl.*, 38, 257–269 (2003).

Masson, I. G. Oberender, A., and Brooking, A. K. "Source separation and potential re-use of re-source residuals at a university campus." *Resour. Conserv. Recycl.*, 40, 155–172 (2004).

McKenzie-Mohr, D. and Smith, W. *Fostering sustainable behavior: An introduction to community based social marketing.* New Society Publishers, Gabriola Island B.C., Canada (1999).

Ojeda-Benítez, S., Armijo-de Vega, C., and Ramírez-Barreto, M. E. "The potential for recycling household waste: A case study from Mexicali, Mexico." *Environ. Urban.*, 12, 163–173 (2000).

Ramos, T. "Development of regional sustainability indicators and the role of academia in this pro-cess: The Portuguese practice." *J. Cleaner Prod.*, 17, 1101–1115 (2007).

Schriberg, M. "Toward sustainable management: The University of Michigan Housing Division's approach." *J. Cleaner Prod.*, 10, 41–45 (2002).

SEDESOL. *Situación actual del manejo integral de los residuos sólidos en México.* Sancho Cer-vera, Mexico City (1999).

Shrum, L. J., Lowrey, T. M., and McCarty, J. A. "Recycling as a marketing problem: A framework fos strategy development." *Psychol. Market.*, 11, 393–416 (1994).

Tabanico, J. and Schultz, P. W. "People aspect of recycling programs: Community based social marketing." *BioCycle*, 41–44 (2007).

Taylor, S. and Todd, P. "An integrated model of waste management behaviour: A test of household recycling and composting intentions." *Environ. Behav.*, 27, 603–630 (1995).

Thøgersen, J. Facilitating recycling: Reverse distribution channel design for participation and sup-port. *Social Marketing Quarterly* 4, 42–55 (1997).

Velázquez, L., Munguia, N., Platt, A., and Taddei, J. "Sustainable university: What can be the mat-ter?" *J. Cleaner Prod.*, 14, 810–819 (2006).

Werder, O. "Influences on the recycling behavior of young adults: Avenues for social marketing campaigns," *Environmental Communication Yearbook.* Lawrence Earlbaum Associates, New York, pp. 77–96 (2005).

Zilahy, G. and Huisingh, D. "The Roles of Academia in Regional Sustainability Initiatives." *J. Cleaner Prod.*, 17, 1057–1066 (2009).

3 Taking Out the Trash (and the Recyclables): RFID and the Handling of Municipal Solid Waste

David C. Wyld

CONTENTS

3.1   INTRODUCTION

This chapter examines how radio frequency identification (RFID) is poised to help transform the way we handle our trash—our municipal solid waste (MSW). We provide an overview showing that trash trends in the United States are not good, as modern life has meant increasing volumes of trash that can be disposed of in fewer and fewer landfills. We examine how RFID can be employed in the MSW area to both facilitate the growth of pay as you throw (PAYT) use-based billing for waste management services and to promote incentive-based recycling programs, both of which aim to reduce the amount of trash entering our landfills. We discuss the prospects for the future as RFID is introduced into what is now $52 billion market for waste handling.

Modern life has become much more complicated and trashy. Every empty coffee cup, box of cereal, tissue, cracked CD case, and so on adds-up. In fact, according to the most recent data available from the U.S. Environmental Protection Agency (EPA), every American man, woman, and child produces—on average—in excess of four and a half pounds of trash (formally referred to as municipal solid waste (MSW)). As can be seen in Figure 3.1, this represents over 75% increase in the per capita amount generated in 1960 and 50% increase over that found in 1980 (U.S. Environmental Protection Agency, 2008).

While the per capita rate has somewhat stabilized over the past two decades, the problem is that with an ever increasing population, the cumulative volume of MSW is rapidly expanding. As can be seen in Figure 3.2, Americans produce a staggering 254 billion tons of trash each year. This represents an approximate 300% increase over the past 50 years (U.S. Environmental Protection Agency, 2008) and, to complicate matters even further, due to a wide range of economic, political and environmental factors, the number of landfills for all this "stuff" to be deposited into has markedly declined. In fact, as can be seen in Figure 3.3, according to the EPA (U.S. Environmental Protection Agency, 2008), today there are less than a quarter of the total number of landfills than were available in the U.S. just two decades ago—down from just under 8,000 in 1988 to 1,754 in 2007. The shortage of landfill space is contributing to an escalation in "tipping fees"—the fees landfills charge to receive a ton of MSW—which range between $10 and $30 per ton in most parts of the country (Wang, 2008). There are already severe shortages of landfill space in pockets of the country. In fact, six states—Alaska, Connecticut, Delaware, North Carolina, New Hampshire, and Rhode Island—have less than 5 years of landfill capacity remaining (National Solid Wastes Management Association, 2008). In these states, and throughout the North East part of the United States, tipping fees have crept much higher, ranging today between $45 and $85 per ton (Abbott, 2008).

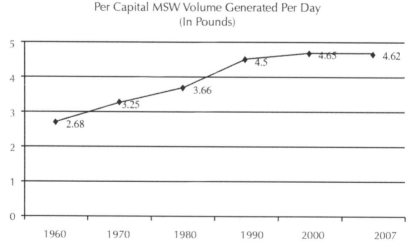

Per Capital MSW Volume Generated Per Day
(In Pounds)

FIGURE 3.1 The rise in individual trash generation, 1960–2007 (U.S. Environmental Protection Agency, 2008).

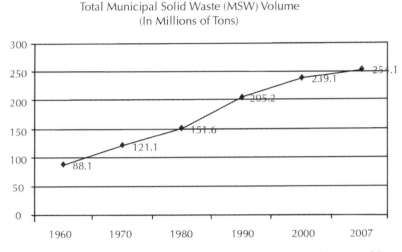

FIGURE 3.2   The rise in overall trash generation, 1960–2007 (U.S. Environmental Protection Agency, 2008).

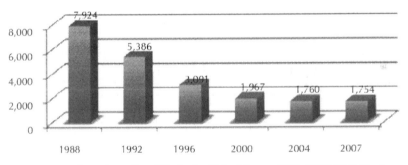

FIGURE 3.3   Total number of landfills in the United States, 1988–2007 (U.S. Environmental Protection Agency, 2008).

Undoubtedly, the business of "trash"–or MSW—is an increasingly important one. It is also an exceedingly complex business, as firms engaged in handling MSW must comply with panoply of environmental rules and regulations, which adds significantly to their operating costs (Portney and Stavins, 2000). Furthermore, there is actually—for lack of a better term—a "trash reverse supply chain" that begins when we place our household waste in a garbage bag, can, or dumpster. Our trash is joined with that of other households and apartment dwellers in the local hauling trash trucks we see on our streets. Yet, with local landfills either being closed or fast reaching their capacities, today it is increasingly common that the trash we throw out at our curbside will be loaded onto larger trucks and offloaded at transfer stations, perhaps several times, before reaching its final resting place at one of the increasingly large "superlandfills"

(Chee, 2008). All of this means that the business of handling, transporting, and processing MSW is becoming a more complex and more expensive logistical operation (World Wildlife Fund, 2009), and all signs point to an increasingly difficult operating environment for waste management companies. Less trash to handle would significantly help the proposition.

While the trash business is an area that many would perceive as a stodgy, low-tech, low-growth business, it is one where RFID presents some intriguing possibilities for waste management. This chapter first provides an overview of RFID technology. Then, we analyze how RFID can reinvent the business model for waste handling through innovative applications of auto-ID technology, revolutionizing the way municipalities and contractors bill for trash collection, and in the process, the manner in which all of us regard "trash". In the process, RFID holds the potential for dramatically reducing the volume of trash and increasing the amount of material being recycled. In the latter regard, RFID can—for the first time—offer real incentives for individuals to participate in recycling programs from their own homes, helping the environment and their communities—and their pocketbooks as well.

3.2   RFID 101

3.2.1   Automatic Identification

Automatic Identification, or Auto-ID, represents a broad category of technologies that are used to help machines identify objects, humans, or animals. As such, it is often referred to as automatic data capture, as auto-ID is a means of identifying items and gathering data on them without human intervention or data entry. Like the omnipresent bar code, RFID is fundamentally another form of auto-ID technology—"a wireless link to identify people or objects" (d'Hont, 2003). The RFID is thus, in reality, a subset of the larger radio frequency (RF) market, with the wider market encompassing an array of RF technologies, including:

- cellular phones,
- digital radio,
- the Global Positioning System (GPS),
- High-definition television (HDTV), and
- wireless networks (Malone, 2004).

The RFID is by no means a "new" technology—as it dates back to World War II (Wyld, 2005). In fact, it is a technology that already surrounds us. First off, if you have an automobile that was manufactured after 1994, the car uses RFID to verify that it is your key in the ignition. Otherwise, the car would not start. If you have an Exxon/Mobil SpeedPass™ in your pocket, you are using RFID. If you have a toll tag on your car, you are using RFID. If you have checked out a library book, you have likely encountered RFID. If you have been shopping in a department store or an electronics retailer, you have most certainly encountered RFID in the form of an Electronic Article Surveillance (EAS) tag.

3.2.2 RFID and Bar Codes

Conceptually, bar codes and RFID are indeed quite similar, as both are auto-ID technologies intended to provide rapid and reliable item identification and tracking capabilities. The primary difference between the two technologies is the way in which they "read" objects. With bar coding, the reading device scans a printed label with optical laser or imaging technology. However, with RFID, the reading device scans, or interrogates, a tag using radio frequency signals.

The specific differences between bar code technology and RFID are summarized in Table 3.1. In summary however, there are five primary advantages that RFID has over bar codes. These are:

1. Each RFID tag can have a unique code that ultimately allows every tagged item to be individually accounted for,
2. RFID allows for information to be read by radio waves from a tag, without requiring line of sight scanning or human intervention,
3. RFID allows for virtually simultaneous and instantaneous reading of multiple tags,
4. RFID tags can hold far greater amounts of information, which can be updated, and
5. RFID tags are far more durable (Wyld, 2005).

TABLE 3.1 RFID and bar codes compared

Bar Code Technology	RFID Technology
Bar codes require line of sight to be read	RFID tags can be read or updated without line of sight
Bar codes can only be read individually	Multiple RFID tags can be read simultaneously
Bar codes cannot be read if they become dirty or damaged	RFID tags are able to cope with harsh and dirty environments
Bar codes must be visible to be logged	RFID tags are ultra thin and can be printed on a label, and they can be read even when concealed within an item
Bar codes can only identify the type of item	RFID tags can identify a specific item
Bar code information cannot be updated	Electronic information can be overwritten repeatedly on RFID tags
Bar codes must be manually tracked for item identification, making human error an issue	RFID tags can be automatically tracked, eliminating human error

3.2.3 How RFID Works

There are three necessary elements for an RFID system to work. These are tags, readers, and the software necessary to link the RFID components to a larger information processing system. In brief, the science of a passive RFID system works like this. The RFID tag is the unique identifier for the item it is attached to. The reader sends out electromagnetic waves, and a magnetic field is formed when the signal from the reader "couples" with the tag's antenna. The unpowered RFID tag draws its power from this magnetic field, and it is this power that enables the tag to send back an identifying response to the query of the RFID reader. When the power to the silicon chip on the tag meets the minimum voltage threshold required to "turn it on", the tag then can respond to the reader through the same RF wave. The reader then converts the tag's response into digital data, which the reader then sends on to the information processing system to be used in management applications. Writing in *Wired*, Singel (2004) likened passive RFID to a "high-tech version of the children's game "Marco Polo" (n.p.). In a passive RFID system, the reader sends out a signal on a designated frequency, querying if any tags are present in its read filed (the equivalent of yelling out "Marco" in a swimming pool). If a chip is present, the tag takes the radio energy sent-out by the reader to power-it-up and respond with the electronic equivalent of kids yelling "Polo" when they are found.

All of this happens almost instantaneously. In fact, today's RFID readers are capable of reading tags at a rate of up to 1,000 tags per second. Through a process known as "simultaneous identification," most RFID systems can capture data from many tags within range of the reader's antenna almost simultaneously. In reality however, the tags are responding individually—within milliseconds of one another—in a manner to prevent tag and reader collision in their signals through response protocols (Wyld, 2005).

3.2.4 Analysis

While it will take a few years for RFID to become commonplace on retail store shelves and the store of the future to become a reality, RFID is already being used in a wide variety of creative applications, including:

- A worker at a distribution center can instantly identify each and every one of the items contained in every box on a pallet on the tongs of the forklift she is driving;
- A librarian can locate a book that had been hopelessly misshelved;
- A worker at a livestock processing facility can instantly access the identity and history of a cow;
- A hospital can locate critical medical devices instantly, wherever they are located throughout the facility;
- A blood bank can track its inventory with greater accuracy;
- A pharmacist can tell that two bottles in his supply of a high in demand, highly addictive prescription drug are counterfeit;
- A military contractor can instantly locate the necessary spare to repair a Blackhawk helicopter;
- An art museum can use RFID-enabled exhibits to provide enhanced visitor experiences by making exhibits come "alive"; and yes,
- A golfer can instantly locate his errant shot and retrieve the ball from the thicket where it landed.

Futurist Paul Saffo foresees that much of the focus on RFID today is on doing old things in new ways, but the truly exciting proposition is the new ideas and new ways of doing things that will come from RFID. He predicts that: "RFID will make possible new companies that do things we don't even dream about" (Van, 2005). As such, this new, old technology will become one of the driving forces of the 21st century. The

RFID is thus an exciting technology, one that is poised to enter our lives in many exciting ways over the next decade. The ability of RFID to deliver rich information, instantaneously and automatically, is why major retailers in the U.S. and abroad, including Wal-Mart, Target, Metro, and Tesco, along with the U.S. Department of Defense, are major backers of employing the technology in their supply chains (Wyld, 2007). And, while much of the media and investment focus has been on such warehousing and retailing applications, now, there is increased interest in applying RFID in a wide variety of settings, including health care (Wyld, 2006a, 2008a, 2008b), sports and entertainment (Wyld, 2006b), museums and theme parks (Wyld, 2006c), and casinos (Wyld, 2008c).

3.3   RFID AND WASTE/RECYCLING

3.3.1   The Municipal Solid Waste Marketplace

Traditionally in the United States, trash collection has been a service performed by municipal Governments—for a flat fee—for its citizens (Canterbury, 1996). Today, cities largely contract out for the service, leading to the rise of several large national firms that dominate the America market, including Waste Management, Allied Waste, BFI, and Republic Services, as well as myriad small local firms that compete as well in this $52 billion annual marketplace (Tracy, 2008). Due to the necessity for such services and the steady cash flow from the monthly billing in this fixed price business model, trash collection is a financially steady and attractive—if sometimes smelly—market for waste management service providers.

However, the single rate model has been criticized not just by environmentalists, but by the Federal EPA as well. The flat rate system provides no incentive for individuals to reduce the amount of waste they put out for collection. As such, heavy users pay the same as light users, making it not only inequitable, but actually harmful to the environment. This is because the flat rate pricing provides no incentive for individuals to participate in recycling programs, encourage composting, or to choose to use source reduction products and packaging (U.S. Environmental Protection Agency, 2008a). In response, some communities have went to hybrid models, charging citizens a flat base rate for a single trash container and then charging a variable rate for additional garbage collection (U.S. Conference of Mayors, 1994), much akin to the model being pursued today with airlines charging more for a second, third, fourth, and so on bag (Sorensen, 2008). Research has shown that some economic disincentives impact individual trash behavior by influencing their cost-benefit calculus by making more trash more expensive (Thogersen, 1994).

3.3.2   Pay As You Throw

There is growing support for a radically different pricing model in the trash business today, known as "Pay As You Throw" (PAYT). Under the PAYT model, people pay a variable rate, based on the amount of trash they actually put out to be collected by the waste management contractor (U.S. Environmental Protection Agency, 2008a). Over 60,000 American cities, as well as many cities across Europe and Australia, currently have PAYT systems. In fact, some have been in place for decades (U.S. Environmental Protection Agency, 1997). However, in the past, such systems have been based on homeowners buying stickers for each garbage can or purchasing specially authorized and/or labeled trash bags, "paying" for each container in which they could "throw" their trash away (Moriarty, 1994). Such long-standing PAYT systems have not gone without issues, including residents intentionally depositing their trash in other people's containers (to avoid their own charges) and a limited rise in the illegal dumping or burning of trash in remote areas (Harder and Knox, 1992). It has also brought about what industry experts have termed the "Seattle stomp" phenomenon. This trend was labeled as such because residents in Seattle, Washington, and other unit pricing cities commonly compact their trash, trying to beat the per-container pricing system by compacting huge amounts of trash into a single can or bag (reducing their trash output by volume, but not by weight) (Villa and Chua, 2009). All in all however, PAYT has been shown to have an impact on households' "trash behavior", significantly decreasing trash output by both weight and volume, while increasing the portion of their waste that was diverted to recycling (Fullerton and Kinnaman, 1996).

Now, RFID technology is being introduced into the waste management industry, making the PAYT model workable. Texas Instruments (TI) has been a leading proponent of using auto-ID technology to not just better the business intelligence of waste management contractors (enabling them to monitor their fleets and worker performance, both for optimizing routing and quality assurance, especially when combined with GPS that is already in wide use in the industry) (Wyatt, 2008). The TI has also demonstrated the workability of PAYT in the field. The key is RFID enabling individual trash containers. Specially-equipped garbage trucks can then weigh each "smart" trash can upon collection, making it possible to ascertain the "net amount" of garbage collected from each customer each time each customer's trash is gathered. The collection process can remain unchanged from what it is today, as the weighing is done when the can is lifted and emptied into the trash truck by the operator, thereby not slowing down the present system performance (Wyatt, Nov. 2008). The TI tests have made use of low frequency RFID tags, due to the harsh environment and the omnipresence of both water (in the content of MSW) and metal (in the trash truck and with metal trash cans in many instances) (Wyatt, 2008). Further, in many urban and even suburban settings, such as apartment buildings, multiple trash cans are in close enough proximity where there would be great potential for misreads and tag collision/confusion.

Whether or not RFID-enhanced PAYT would prove to be revenue enhancing, neutral or negative overall for cities and their waste management contractors remains to be seen. The accuracy possible through the use of automatic identification technology does make possible new concepts for individual accountability and tracking. However, the PAYT concept certainly encourages more individual environmental responsibility when it comes to household management of MSW. The one thing that is assured is that it does encourage folks to recycle what can recycled from their own trash, decreasing their net trash output and thus, their weight-based trash charges. With RFID making it more possible to accurately assess weight and volume-based trash charges for each customer, this will yield more recycling incentives than ever. And now, RFID is being brought to bear to directly encourage recycling through tracking and "incentivizing" the process for individuals.

3.3.3 Growing Recycling

According to the most recent data available (for the 2007 calendar year), the EPA found that just over a third of all MSW in the United States is actually recycled. With only about 12% of all MSW is burned for energy recovery or simply incinerated, this means that over half of our total trash output—54%—still ends-up simply reposited into ever-fewer landfills (U.S. Environmental Protection Agency, 2008). Paper and paperboard is the largest category of our trash output, comprising almost a third of the total. Yet, as can be seen in Figure 3.4, barely half (54.5%) of our paper products are actually recycled. In fact, the EPA data shows that recycling rates overall lag expectations (U.S. Environmental Protection Agency, 2008).

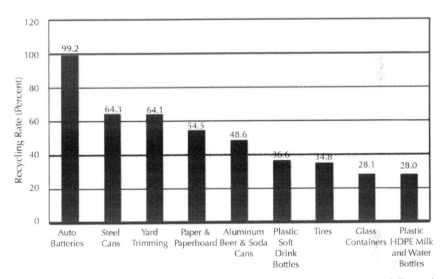

FIGURE 3.4 Recycling rates of selected products, 2007 (U.S. Environmental Protection Agency, 2008).

Why does participation in recycling efforts lag? Analysts often point to cumbersome recycling requirements imposed by cities and their waste contractors (Lansana, 2005). Such program requirements require citizens to not just separate their recyclables by product category, but ask them to put specific items out for pick-up on specific days (i.e., glass on Mondays, paper on Wednesdays, plastics on Fridays) or to take the items to recycling collection centers, rather than setting the items out with their "normal" trash on their "normal" collection days.

Today, innovative recycling solutions providers are looking to use RFID to make recycling "easier" and to track the recycling patterns of individual households. Some are even finding a way to "incentivize" individuals into recycling behavior by not just reducing their PAYT garbage bills, but actually paying or rebating them directly for the amount of recyclabes they divert from the landfill. There are several firms vying for this market, including RecycleBank (http://www.recyclebank.com/), based in New York City, Routeware (http://www.routeware.com/), based in Beaverton, Oregon, and an Irish firm, Advanced Manufacturing Control Systems (AMCS) (http://www.amcs.ie/). Austin Ryan, cofounder and business development director for AMCS, recently commented that: "Increasing recycling rates requires the deployment of creative new strategies and technologies in the waste management industry" (Anonymous, 2008). Each of these firms are marketing solutions whereby the recycling collection process makes use of special RFID-tagged recycling containers (using low-frequency RFID tags), which are collected by trucks equipped with smart scales that read the tags (to associate the collection with a particular customer) and to weigh that customer's recyclables (based on the weight of the filled container versus the empty container weight) (Anonymous, 2008; O'Connor, 2008).

For example, RecycleBank currently serves a number of cities—(the largest of which is Philadelphia) in the North East, covering several hundred thousand homes. RecycleBank's system works in tandem with existing municipal waste management contractors' collections, as they do not operate their own collection equipment. They do provide customers with RFID equipped recycling carts, ranging between 35 to 96 gallons in size. In these bins, residents pour all recyclable materials. Once collected by RFID-equipped collection trucks, the customer's account is credited for the weight of the contents in the cart (Swedberg, 2008). The amount of material recycled is converted into RecycleBank Points, which can use at participating reward partners. These include national and local retailers such as:

- Bed, Bath and Beyond
- CVS/pharmacy
- Dick's Sporting Goods
- Foot Locker
- Home Depot
- Petco.com
- Rite-Aid
- Sears
- Starbucks
- Target.com (Crowe, 2009).

The recyclable materials—paper, plastics, cardboard, aluminum, and so on—are then separated at processing centers. After being separated by type, the material can then be directed towards reuse (Crowe, 2009).

What are the results? Ron Gonen, RecycleBank's cofounder and CEO, reports that the benefits of incentivizing the recycling behaviors of individuals can make whole cities much greener. In fact, Gonen reports that: "We've taken cities with almost no recycling and brought them to 40% of their trash being diverted from waste" (O'Connor, 2008). For municipalities and waste haulers, this means that rather than having to pay the rising tipping fees for delivering MSW to landfills, they can actually reverse the equation, earning money on the volume of waste products that are directed towards recycling (Abbott, 2008). For the customer, RecycleBank provides incentive credits based on their actual recycling volume, offering discounts and credits at hundreds of retail partners, ranging from the national brands (listed previously) to local retailers, restaurants, and grocery stores. Kraft Foods is one of the lead sponsors of RecycleBank, offering discounts on its family of products as incentives for consumer recycling. Kraft's Elisabeth Wenner, the firm's director of sustainability, says that the value proposition for her company is that by encouraging recycling, Kraft helps reduce the amount of its own and others' product packaging in landfills. Thus, according to Wenner, "RecycleBank offers an innovative way to make it easy and rewarding for consumers to recycle" (O'Connor, 2008). For corporate partners, the RecycleBank incentive program offers a marketing tool to encourage both first use of their products or services and to promote repeat transactions. Thus, they are a way of "doing well by doing good", promoting both individual and corporate environmental responsibility——and a unique marketing program at the same time.

3.4 ANALYSIS

All in all, the MSW market holds the potential for rapid development over the next few years for RFID solutions providers, as well as those vendors providing the hardware and software necessary to support PAYT and for monitoring recycling. In fact, today's economic conditions could work to benefit solutions providers in this area by accelerating the growth of both the PATY and recycling incentive programs, both in the U.S. and abroad. This is evidenced by the growing interest in such programs across Europe (Lansana, 2005; O'Connor, 2008). Likewise, the concern over the impact of MSW on climate change could also work to spur the growth of both PAYT and greater recycling efforts (U.S. Environmental Protection Agency, 2008b). This is evidenced by the rapid growth of an incentive-based recycling program in Michigan. Introduced by a start-up firm, Rewards for Recycling, the company has enlisted over 80,000 households across several counties in Michigan in just its first 6 months of operations (O'Connor, 2009). Thus, the curbside may be one of the most promising areas for RFID technology to be employed, not just for profits, but for a greener world as well. In industry after industry, RFID has proven to be a transformative, game-changing technology, producing new levels of efficiency, customer service, and business intelligence. We should expect no less in the world of trash, in order to minimize the amount of trash and maximize the health of the planet.

KEYWORDS

- **Business Intelligence**
- **Environment**
- **Municipal solid waste**
- **Recycling**
- **Radio frequency identification technology**
- **Waste Management**

REFERENCES

Abbott, K. *White paper: The economics of recycling (Released September 2008)*. [Online]. Available: http://recyclebank.com/recycling/economics [Accessed: November 20, 2008].

Anonymous. "Irish Company's RFID Waste Management Technology Comes to the United States." *American Recycler*, November 2008. [Online]. Available: http://www.americanrecycler.com/1108/irish.shtml. [Accessed: October 1, 2009].

Canterbury, J. L. "Pay-as-you-throw: The Bonus Beyond Waste Prevention." *Waste Age*, (May 1, 1996). [Online]. Available http://wasteage.com/mag/waste_payasyouthrow_bonus_beyond/. [Accessed: November 16, 2008].

Chee, M. "Tracking the trash: Avatar Partners' RFID solution adds efficiency and accuracy to Allied Waste's Disposal Services." *RFID Product News*, Spring 2008. [Online]. Available: http://www.rfidproductnews.com/issues/2008.06/cover.story.php [Accessed: December 30, 2008].

Crowe, A. "RecycleBank is the eco-friendly way to cash in your trash." *Boston Happenings Examiner* (January 10, 2009). [Online]. Available: http://www.examiner.com/x-1987-BostonHappenings-Examiner~y2009m1d10-RecycleBank-helps-you-cash-in-for-doing-good [Accessed: June 27, 2009].

d'Hont, S. *The cutting edge of RFID technology and applications for manufacturing and distribution: A white paper from Texas Instruments (Released July 2003)*. [Online]. Available: http://www.ti.com/tiris/docs/manuals/whtPapers/manuf_dist.pdf [Accessed: April 23, 2004].

Fullerton, D. and Kinnaman, T. C. "Household responses for pricing garbage by the bag." *Am. Econ. Rev.*, **86**(4), 971–984 (November 1996).

Harder, G. and Knox, L. "Implementing variable trash collection rates. *BioCycle*, **33**(4), 18–31 (April 1992).

Lansana, F. M. "A comparative analysis of curbside recycling behavior in urban and suburban communities." *The Professional Geographer*, **45**(2), 169–179 (June 2005).

Malone, R. "Reconsidering the role of RFID." *Inbound Logistics*, August 2004. [Online]. Available: http://www.inboundlogistics.com/articles/supplychain/sct0804.shtml [Accessed: September 18, 2004].

Moriarty, P. "Financing waste collection for maximum diversion." *BioCycle*, **35**(1), 15–18 (January 1994).

National Solid Wastes Management Association. *White paper: Modern landfills—A far cry from the past*, [Online]. Available: http://wastec.isproductions.net/webmodules/webarticles/articlefiles/463-white%20paper%20landfill%20final.pdf [Accessed: November 15, 2008].

O'Connor, M. C. "Routeware launches RFID solution for waste haulers: The system employs low-frequency RFID interrogators on trash-collection trucks to identify tagged waste and recycling containers, as well as track the recycling efforts of the residents they serve." *RFID Journal*, (May 5, 2008). [Online]. Available: http://www.rfidjournal.com/article/articleprint/4067/-1/1/ [Accessed: February 15, 2009].

O'Connor, M. C. "Michigan households get RFID-enabled rewards for recycling." *RFID Journal* (October 9, 2009). [Online]. Available: http://www.rfidjournal.com/article/view/5293/ [Accessed: October 28, 2009].

Portney, P. and Stavins, R. N. *Public policies for environmental protection*. RFF Press, Washington, DC (2000).

Singel, R. "American passports to get chipped." *Wired* (October 21, 2004). [Online]. Available: http://www.wired.com/news/privacy/0,1848,65412,00.html [Accessed: December 2, 2004].

Sorensen, J. *White paper: More airlines worldwide choosing revenue-based methods to serve and reward customers (Issued December 1, 2008)*. [Online]. Available: http://www.ideaworkscompany.com/press/2008/AnalysisRevenueBased2008.pdf [Accessed: June 17, 2009].

Swedberg, C. "RFID helps reward consumers for recycling: Kraft Foods joins RecycleBank in its use of RFID to track and reward consumers for recycling." *RFID Journal* (February 22, 2008). [Online]. Available: http://www.rfidjournal.com/article/articleprint/3936/-1/1/ [Accessed: June 11, 2009].

Thogersen, J. "Monetary incentives and environmental concern: Effects of a differentiated garbage fee." *J. Consum. Policy*, **17**(4), 407–442 (December 1994).

Tracy, P. "Waste Management (WMI) turns trash into nearly \$1.2 billion in cash a year." *Street Authority Market Advisor*, April 21, 2008. [Online]. Available: http://www.topstockanalysts.com/cmnts/2008/04-21-waste-management.asp [Accessed: July 18, 2008].

U.S. Conference of Mayors, *A primer on variable rate pricing for solid waste services*. The United States Conference of Mayors, Washington, DC (1994).

U.S. Environmental Protection Agency (EPA), *General public—Throw away less and save (Released April 1997)*. [Online]. Available: http://www.epa.gov/osw/conserve/tools/payt/tools/public.htm [Accessed: December 12, 2008].

U.S. Environmental Protection Agency (EPA). *Fact sheet: Climate change and municipal solid Waste (MSW) (Released September 2008b)*. [Online]. Available: http://www.epa.gov/osw/conserve/tools/payt/tools/factfin.htm [Accessed: June 20, 2009].

U.S. Environmental Protection Agency (EPA). *Municipal solid waste generation, recycling, and disposal in the United States: Facts and figures for 2007 (Released November 2008)*. [Online]. Available: http://www.epa.gov/epawaste/nonhaz/municipal/pubs/msw07-fs.pdf [Accessed: December 10, 2008].

U.S. Environmental Protection Agency (EPA). *Pay-as-you-throw (Released September 2008a)*. [Online]. Available: http://www.epa.gov/osw/conserve/tools/payt/ [Accessed: December 12, 2008].

Van, J. "RFID spells media revolution, futurist says." *Chicago Trib.*, **24**(104), B1 (April 16, 2005).

Villa, C. M. and Chua, A. J. "Dirty talk—Paying for trash. PAYT in the Philippines: Issues and concerns." *BusinessWorld Online* (January 1, 2009). [Online]. Available: http://www.bworldonline.com/Features_2008/content.php?id=IDEA010109 [Accessed: August 16, 2009].

Wang, C., Lin, M., and Lin, C. "Factors influencing regional municipal solid waste management strategies." *J. Air Waste Manage. Assoc.*, **58**(7) (July 1, 2008). [Online]. Available: http://www.highbeam.com/doc/1G1-181856762.html [Accessed: January 2, 2009].

World Wildlife Fund. *Municipal Solid Waste Factsheet–2008*. [Online]. Available: http://www.wwfpak.org/factsheets_mswf.php [Accessed: January 13, 2009].

Wyatt, J. "Maximizing waste management efficiency using RFID." *RFID Product News*, Fall 2008. [Online]. Available: http://www.rfidproductnews.com/issues/fall2008/waste.php. [Accessed: September 15, 2009].

Wyatt, J. "RFID waste management: Pay-as-you-throw." *RFID World* (November 10, 2008). [Online]. Available: http://www.rfid-world.com/showArticle.jhtml?articleID=212001679 [Accessed: September 15, 2009].

Wyld, D. C. *RFID: The right frequency for government—A research report from The IBM Center for the Business of Government* (September 2005). [Online]. Available: http://www.businessofgovernment.org/main/publications/grant_reports/details/index.asp?gid=232 [Accessed: October 3, 2005].

Wyld, D. C. "The importance of pedigree: Why instituting RFID-based tracking of pharmaceuticals is essential to counteracting counterfeiting and maintaining both the health of the public and the potency of the American drug industry." *Competition Forum*, **4**(1), 261–266 (October 2006a).

Wyld, D. C. "Sports 2.0: A look at the future of sports in the context of RFID's 'Weird New Media Revolution.'" *Sport J.*, **9**(4), 3–17 (December 2006b).

Wyld, D. C. "DaVinci uncoded: RFID is enhancing the management and experience of art galleries and museums...and even becoming a part of the artists' palettes for creating works of art." *Global Identification*, (26), 36–40, (May 2006c).

Wyld, D. C. "RFID 101: The next big thing for management." *Eng. Manag. Rev.*, **35**(2), 3–19 (May 2007).

Wyld, D. C. "The implant solution: Why RFID is the answer in the highly unique orthopedic supply chain, providing ROI for suppliers and assurance for patients and their surgeons." *ID World*, (15), 12–15 (June 2008a).

Wyld, D. C. "Playing a deadly game of match: How new efforts to use RFID in blood banking and transfusion can save patient lives and safeguard the blood supply chain." *Global Identification*, (37), 24–26 (March 2008b).

Wyld, D. C. "Radio frequency identification: Advanced intelligence for table games in casinos," *Cornell Hospitality Quarterly*, 49(2), 134–144 (June 2008c).

4 Management of Municipal Solid Waste by Vermicompost–A Case Study of Eluru

Sudhir Kumar J., Venkata Subbiah K., and Prasada Rao P.V.V.

CONTENTS

4.1 INTRODUCTION

Solid waste is an unwanted byproduct of modern civilization. Landfills are the most common means of solid waste disposal. But, the increasing amount of solid waste is rapidly filling existing landfills, and new sites are difficult to establish. Alternatives to landfills include the use of source reduction, recycling, composting, and incineration, as well as use of landfills. Incineration is most economical if it includes energy recovery from the waste. Energy can be recovered directly from waste by incineration or the waste can be processed to produce storable refuse derived fuel (RDF). Information on the composition of solid wastes is an important in evaluating alternative equipment needs, systems, management programs, and plans. Household surveys are done in six divisions of Eluru Municipal Corporation, A.P, India and per capita waste for the corporation is estimated. Pulverization of municipal solid waste is done and the pulverized solid waste is dressed to form a bed and the bed is fed by vermi's which converts the bed into vermicompost. The obtained vermicompost is sent to recognized lab for estimating the major nutrients that is potassium (K), phosphorous (P), nitrogen (N), and micronutrient values. It is estimated that 59–65 tons of wet waste is generated in Eluru per day and if this wet waste is converted to quality compost 12.30 tons of vermicompost can be generated. If Municipal Corporation of Eluru (MCE) manages this wet waste, an income of over rupees 0.89 crores per annum can be earned by MCE which is a considerable amount for providing of better services to public.

There has been a significant increase in municipal solid waste (MSW) generation in India in the last few decades. This is largely because of rapid population growth and economic development in the country. Solid waste management has become a major environmental issue in India. The per capita of MSW generated daily, in India ranges from about 100 gm in small towns to 500 gm in large towns[1]. The population of Mumbai grew from around 8.2 million in 1981 to 12.3 million in 1991, registering a growth of around 49%. On the other hand, MSW generated in the city increased from 3–200 tons/day to 5355 tons/day in the same period registering a growth of around 67% (CPCB, 2000)[2]. This clearly indicates that, the growth in MSW in our urban centers has outpaced the population growth in recent years. This trend can be ascribed to our changing lifestyles, food habits, and change in living standards. The MSW in cities is collected by respective municipalities and transported to designated disposal sites, which are normally low lying areas on the outskirts of the city. The limited revenues earmarked for the municipalities make them ill-equipped to provide for high costs involved in the collection, storage, treatment, and proper disposal of MSW. As a result, a substantial part of the MSW generated remains unattended and grows in the heaps at poorly maintained collection centers. The choice of a disposal site also is more a matter of what is available than what is suitable. The average collection efficiency for MSW in Indian cities is about 72.5% and around 70% of the cities lack adequate waste transport capacities (TERI, 1998). The insanitary methods adopted for disposal of solid wastes is, therefore, a serious health concern. The poorly maintained landfill sites are prone to groundwater contamination because of leach ate production. Open dumping of garbage facilitates the breeding for disease vectors such as flies, mosquitoes, cockroaches, rats, and other pests (CPCB, 2000). The municipalities in India therefore face the challenge of reinforcing their available infrastructure for efficient MSW management and ensuring the scientific disposal of MSW by generating enough revenues either from the generators or by identifying activities that generate resources from waste management.

4.1.1 Per Capita Quantity of Municipal Solid Waste in Indian Urban Centers

The quantity of waste from various cities was accurately measured by NEERI. On the basis of quantity transported per trip and the number of trips made per day the daily quantity was determined. The quantity of waste produced is lesser than that in developed countries and is normally observed to vary between 0.2–0.6 kg/capita/day. Value up to 0.6 kg/capita/day is observed in metropolitan cities in the Table 4.1 below. The total waste generation in urban areas in the country is estimated to be around 38 million tons per annum (mtpa)[3]. Forecasting waste quantities in the future is as difficult as it is in predicting changes of waste composition. The factors promoting change in waste composition are equally relevant to changes in waste generation. Storage methods, salvaging activities, exposure to the weather, handling methods, and decomposition, all have their effects on changes in waste density.

TABLE 4.1 Quantity of municipal solid waste in Indian urban centers

Population Range (in million)	Number of Urban Centers (sampled)	Total population (in million)	Average per capita value (kg/capita/day)	Quantity (tons/day)
<0.1	328	68.300	0.21	14343.00
0.10.5	255	56.914	0.21	11952.00
0.51.0	31	21.729	0.25	5432.00
1.02.0	14	17.184	0.27	4640.00
2.05.0	6	20.597	0.35	4640.00
>5	3	26.306	0.50	13153.00

4.1.2 Town Profile

Eluru, previously known as Helapuri and has a rich cultural and political history. It was a part of Buddhist Kingdom called Vengi. During the Chalukyas (700–1200 AD), Eluru was a province. Later on Eluru remained a part of Kalinga Empire. During division of Northern cirkaras into district, Eluru made a part of Machilipatnam district. Later it was included in the Godavari district in 1859. Subsequently, Eluru made part of Krishna district. Finally in the year 1925, West Godavari District was formed with Eluru as its headquarters. Eluru town is situated at 16.7° N latitude and 81.1° E longitude on the Kolkata-Chennai National Highway (NH-5). The Visakhapatnam-Chennai railway line passes through the town. Eluru was a selection grade municipality of Andhra Pradesh. It has been upgraded to Municipal Corporation on 09.04.2005. The area of Eluru Municipal Corporation is 14.55 sq.km with a population of 1,90,062 as per 2001 Census. It would be seen that during the last decade Eluru experienced a negative population growth[4].

4.1.3 MSW Availability

It is understood by household survey, that 70–75 tons of MSW in Eluru is being generated every day. The available quantities can safely and conveniently generate about 3 MWs of power or can be converted in to vermincompost as manure for farmers.

4.1.4 House Hold Survey

To collect the house hold garbage, first of all pick 10% of houses in the division so that the sample shall be correct. In collecting sample, there must include all the type of constructions like schools, colleges, factories, hostels, hospitals, and so on. The samples are collected separately that is wet (vegetable waste, kitchen waste, etc.) and dry waste (papers, room waste, bags, boxes, etc.). This sampling process is continued for seven days so that we can predict the average value. The results of the analyses show that MSW contains organic matter and miscellaneous materials (bricks, fine dust, rubber, wood, leather, wastewater, etc.). The percentage of recyclable materials (glass, paper, plastic, metals) has been found to be very low. This may be due to rag pickers, who collect and segregate recyclable materials from collection points and disposal sites. The results from the survey reveal that the per capita MSW generation rate is nearly 0.12 kg/capita/day. The per capita generation rate for various areas in Eluru city calculated. This rate varies from 0.14 kg/capita/day in Division 23 to 0.09 kg/capita/day in Division 50 where as 0.6 kg /capita/day generation of MSW observed in metro cities. The households are selected randomly from the divisions so that the entire area of the division is covered. The opinion of the public regarding the services of MCE collected from the questionnaires.

4.2   WHAT IS VERMICULTURE

Vermicompost is an organic manure (biofertilize) produced as the vermicast by earth worm feeding on biological waste material; plant residue. Earthworms are small, soft, cylindrical bodied invertebrates that play a vital role in soil ecosystem maintenance. Earthworms greatly influence soil properties and cast production, which results in the continuous turnover of the soil and mixing of minerals and organic constituents. Worms that live in the soil are the farmer' and gardener's friends. Vermicompost, the end product, is extremely useful for enriching and fertilizing the soil. It is odorless and safe to handle. It is rich in hormones, antibiotics, and vitamins that produce healthy plant growth. Although, its nitrogen, phosphorus, and potassium values are not as high as for chemical fertilizers, it is multipurpose compost that provides all the ingredients needed to improve most soils and is much better for the environment, as well. Vermi-compost is also seven times richer than compost that has been rotted without introduc-tion of worms, so only one seventh of the quantity is needed to enrich the soil. Tests in India have shown that vermincompost application can double wheat yields and quadruple yields of fodder. For best effect vermincompost needs to applied before the growing season over a 2 or 3 year period[5].

TABLE 4.2   Nutrient content of vermicompost

Nitrogen	0.8 to 1.0 %
Phosphorous	0.8 to 1.0 %
Potash	0.8 to 1.0 %
Calcium	0.44%
Magnesium	0.15%
Iron	27.3 ppm
Manganese	16.4 ppm
Zinc	18.0 ppm
Copper	7.6 ppm

TABLE 4.3 Dosage of compost

Field crops	2 tons/ha
Horticulture	200 gm/plant (Young)
Crops	5 kg/tree (Matured)
Forest	200 gm/plant (Young)
Ornamental	50 gm/pot

4.2.1 Method of Preparation of Vermicompost

A thatched roof shed preferably open from all sides with unpaved (katcha) floor is erected in EastWest direction length wise to protect the site from direct sunlight. A shed area of 12′×12′ is sufficient to accommodate three vermin beds of 10′×3′ each having 1′ space in between for treatment of 9–12 quintals of waste in a cycle of 40–45 days. The length of shed can be increased/decreased depending upon the quantity of waste to be treated and availability of space. The height of thatched roof is kept at 8 ft from the center and 6 ft from the sides. The base of the site is raised at least 6 inches above ground to protect it from flooding during the rains. The vermin beds are laid over the raised ground as per the procedure given below[6]. The site marked for vermin beds on the raised ground is watered and a 4″–6″ layer of any slowly biodegradable agricultural residue such as dried leaves/straw/sugarcane trash etc. is laid over it after soaking with water. This is followed by 1″ layer of vermicompost or farm yard manure. The loaded waste is finally covered with a jute mat to protect earthworms from birds and insects. Water is sprinkled on the vermin beds daily according to requirement and season to keep them moist. The waste is turned upside down fortnightly without disturbing the basal layer (vermin bed).The appearance of black granular crumbly powder on top of vermin beds indicate harvest stage of the compost. Watering is stopped for at least 5 days at this stage. The earthworms go down and the compost is collected from the top without disturbing the lower layers (vermin bed). The first lot of Vermicompost is ready for harvesting after 2–2 ½ months and the subsequent lots can be harvested after every 6 weeks of loading. The vermin bed is loaded for the next treatment cycle. A tractor load of MSW is collected and it is dumped in the dump yard.

- The MSW is segregated that is all the dry wastes such as clothes, carry bags, and other dry wastes are segregated from wet waste.
- The wet waste is pulverized and arranged in the form of bed of dimensions as per the table below.

TABLE 4.4 Dimensions of the vermi bed

Length	60 inches
Width	49 inches
Height	23 inches

- The bed is wetted and the compost is prepared according to the above procedure shown in Figures 4.1–4.2.
- There will no changes in the dimensions of bed and in the weight of MSW when it changes to vermin compost.

4.2.2 Multiplication of Worms in Large Scale

Prepare a mixture of cow dung and dried leaves in 1:1 proportion. Release earthworm at the rate 50 numbers/10 kg. Of mixture and mix dried grass per leaves or husk and keep it in shade. Sprinkle water over it, time to time, to maintain moisture level. By this process, earthworms multiply 300 times within 1–2 months. These earthworms can be used to prepare vermin compost.

FIGURE 4.1   (a) Pulverization, (b) Segregation, (c) Separation, (d) Preparation of Vermi bed

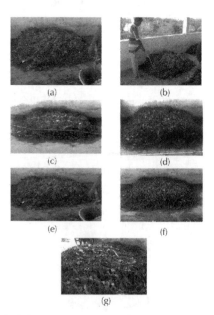

FIGURE 4.2   (a) Vermi bed (b) Watering of bed, (c), (d), (e), (f) (g) Images of gradual change of MSW to vermicompost.

TABLE 4.5 Nutrient values of prepared vermicompost by Lab analysis

Sl. No	Tests	Units of measurement	Results obtained
1	Nitrogen (as N)	% by mass	1.02
2	Phosphorus (as P)	% by mass	0.13
3	Potassium (as K)	% by mass	0.27
4	Magnesium (as Mg)	% by mass	0.06
5	Zinc (as Zn)	% by mass	0.010
6	Boron (as B)	% by mass	<0.001
7	Copper (as Cu)	% by mass	0.003
8	Iron (as Fe)	% by mass	0.41
9	Manganese (as Mn)	% by mass	0.03

4.3 APPLICATIONS OF VERMICOMPOST

Add 15–20 gm of vermicompost to a liter of water and use this to water potted plants daily. Use 1 part of vermi castings, 1 part sand, and 1 part garden soil and mix well before use. Sprinkle vermincompost on surface of the soil and water as usual. Repeat every 40–50 days. Prepare the nursery bed, mix vermi castings, with top soil (1 kg/sq. m) plant and water the grass. For crop like paddy, ragi and legumes, sugercane, cotton, vegetables, and so on, apply 300–500 kg/acre by broadcasting. Apply 1–3 kg/tree (depending upon age) twice a year. For crops like coconut, rubber, groundnut, mango, cashew, and other plantation crops like:

TABLE 4.6 Application of vermicompost for different fields

Banana	1 metric ton/acre
Flowers	2 metric ton/acre
Grapes	1.5 metric ton/acre
Tea	1.5 metric ton/acre
Coffee	1 metric ton/acre
Mulberry	1 metric ton/acre

TABLE 4.7 Vermicast *vs.* chemical fertilizers in soil

Criteria for comparison	Chemical fertilizers	Vermicast
Macro nutrient contents	Mostly contains only one (N in urea) or at the most two (N & P in DAP) nutrients in any one type of chemical fertilizer	Contains all i.e nitrogen (N), phosphorus (P) & potassium (K) in sufficient quantities
Secondary nutrient contents	Not available	Calcium (Ca), magnesium (Mg) & sulphur (S) is available in required quantities
Micro nutrient contents	Not available	Zinc (Zn), boron (B), manganese (Mn), iron (Fe), copper (Cu),molybdenum (Mo) and Chlorine (Cl) also present
pH balancing	Disturb soil pH to create salinity and alkalinity conditions	Helps in the control of soil pH and checks the salinity and alkalinity in soil
EC errection	Creates imbalance in soil EC affecting nutrients assimilation	Helps in balancing the EC to improve plant Nutrient adsorption the salinity and alkalinity in soil

TABLE 4.7   *(Continued)*

Criteria for comparison	Chemical fertilizers	Vermicast
Organic carbon	Not available	Very high organic carbon and umus Contents improves soil characteristics
Moisture retention capacity	Reduces moisture retention capacity of the soil	Increases Moistures retention capacity of the soil
Soil Texture	Damages solid texture to reduce aeration	Improves soil texture for better aeration
Beneficial bacteria & fugi	Reduces Biological activities and thus the Fertility is impaired	Very high biological life improves the soil Fertility and productivity on sustainable basis
Plant growth hormones	Not available	Sufficient quantity helps in better growth And production

4.3.1   Total Vermicompost that can be Obtained from Eluru Town

Total waste generated in Eluru per day	58.55 tons
4560% of food and garden waste is available in total waste for low income cities	
Total food and garden waste available per day	58.55*0.525
	30.74 tons
Compost obtained from the solid waste	30.74*0.4
	12.29 tons

4.3.2   Cost Analysis

Estimated cost of 0.001 ton of compost	Rs. 2/
Estimated cost of 12.3 tons of compost	Rs. 24,593
Income that can earn by MCE per day	Rs. 24,593
Income that can earn by MCE per anum	Rs. 0.89 crores

TABLE 4.8   Wages for workers engaged in the collection of MSW

SL.NO	POST	WAGES PER ANNUM in Rs
1	Senior permanent workers	1,08,000
2	Junior permanent workers	84,000
3	Dokras	46,800

4.3.3   Expenditure of MCE in the Collection of MSW

Senior permanent workers	Rs. 1,08,000*100	Rs. 1.08 Crores per annum
Junior permanent workers	Rs. 84,000*250	Rs. 0.67 Crores per annum
Dokras	Rs. 46,800*145	Rs. 0.06 Crores per annum
Fuel	64 Liters*33 Rs*300 Day	Rs. 0.06 Crores per annum
Total expenditure per Annum		Rs. 3.91 Crores

Therefore, by managing the wet waste generated in the town, MCE can earn Rs 0.89 crores per annum which is a considerable amount for providing better services.

4.4 CONCLUSION

The municipal corporations being the responsible authority in India for MSW in addition to wide range of responsibilities related to health and sanitation, have not been very effective as far as MSW services are concerned. Collection, transportation, and disposal of all the three components of waste lack in terms of infrastructure and maintenance up-gradation however, the weakest link in the chain of waste management in Indian situations is the collection of waste.

This analysis unambiguously shows that recycling impact is of importance in the prediction of solid waste generation. The degree of accuracy of this model is determined by the reliability of the published information, which has been provided by MCE.

Experience indicates the estimation of solid waste generation is crucial for the subsequent system planning of solid waste management in the metropolitan and rural regions from both short and long term perspective. However, a complete record of solid waste generation and composition is not always present. The central idea of vermicompost is not only to manage the solid waste system by producing wealth from it but also to save the environment from pollution.

KEYWORDS

- **Garbage**
- **Landfill**
- **Municipal solid waste**
- **Municipal Corporation of Eluru**
- **Vermicompost**

NOTES

[1] Singhal, S. and Pandey, S. Solid Waste Management in India: Status and Future Directions, pp 234–241.

[2] en.wikipedia.org/wiki/Municipal_solid_waste (accessed on June 14, 2010).

[3] www.hpurbandevolpopment.nic.in/swm/chap3.pdf

[4] Data collected from Municipal Corporation of Eluru.

[5] www.agricare.org (accessed on June 14, 2010).

[6] www.agri.and.nic.in/vermi_culture (accessed on June 14, 2010).

5 Waste Management Program at the Universidad Tecnologica de Leon

Dolores Elizabeth Turcott Cervantes, Karina Guadalupe López Romo, and Mario Bernardo Reyes Marroquín

CONTENTS

5.1   INTRODUCTION

In Mexico a lot of educational institutions send their waste to places of final disposal, which generates a negative impact to the environment; especially when these places are not adequate and they do not comply with the current environmental legislation. This is why, at the Universidad Tecnologica de Leon (UTL) it has been implemented a System of Environmental Management; where 83% of the negative impact (environmental aspects) is related to waste generation. This resulted in the creation, and put into practice, of a Waste Management Program.

Different actions have taken place to make the program work. For example, creating a plan for the handling of valued waste, design and set forth of the infrastructure for the primary separation of waste, environmental education, and promotion to the university's community about the adequate handling of the waste, among other things.

Also, one basic part of the program is the creation of indicators in 2008, 2009, and 2010, the daily total generation of waste, per capita generation, the amount recovered in the storage center for its sale and eventual recycling as the waste used to elaborate natural fertilizer (compost). Therefore, in this chapter are shown the results obtained from the creation and implementing of the Management Waste Program of the UTL, which can be used as testimony and model to continue bettering the handling of waste inside educational institutions.

All activities inside the university campus cause, in certain degree, a negative impact to the environment. One of these impacts is the generation of waste. It is important that all educational institutions, mainly universities, to implement actions to not only support the caring of the environment, but also to contribute to the overall formation of the students. So they, the students, can be more prepared for challenges in the near future.

In Mexico, there are universities that have Waste Management Programs, some of these examples are: UAM (Universidad Autonoma Metropolitana), UNAM (Universidad Nacional Autonoma de Mexico), Tecnologico de Monterrey (just some of their campus), Universidad Autonoma del Estado de Morelos, Universidad de Guadalajara, Universidad Autonoma de Baja California, Universidad Autonoma de San Luis Potosi, Instituto Tecnologico Autonomo de Mexico, Escuela de Estudios Superiores de Zaragoza. In Guanajuato, ITESI (Instituto Tecnologico Superior de Irapuato, which has an ISO 14001 certification), Universidad de Guanajuato, Universidad Iberoamericana de Leon, and Tecnologico de Monterrey Campus Leon (both are starting their programs), and UTL.

In the UTL, there has always been a concern for addressing and minimizing this negative impact. We have been the main promoters of this change: teachers and students of the Environmental Technology degree. In 2008, it has been said that the UTL was the only educational institution in Leon, Guanajuato (Mexico) which was closest to the concept of "a green university" (Palacios, 2008).

Although, the degree of Environmental Technology opened in 1998 and since then a lot of actions have been taken to protect the environment, it was not until 2006 that nine students did their evaluations to create a system of environmental management inside the university (named: SGA-UTL), with the purpose of formalizing and integrating the efforts done in the past and formulating significant and non-significant environmental aspects; taking as reference the ISO 14001.

Out of the 18 environmental aspects that were identified for the SGA-UTL, approximately 83% correspond to the impact caused by the UTL in the area of waste (Estrada et al., 2007; Reyes, 2008). So, to follow-up and respond to the SGA-UTL, in 2008 started in a formal way, the Waste Management Program in conjunction with the operation of a storage center, which helps with the collection, storage, and separation of the waste. The inorganic waste recovered is sold for later recycling and the organic waste is used to elaborate natural fertilizer (compost).

Just in 2008, the UTL generated 55.77 tons of waste (on average 0.2 tons per day), with this, we can compare ourselves to other universities in Mexico; for example, in the Universidad Autonoma de Baja California (Mexicali I campus) one ton is generated a day (Ojeda and Ramírez, 2007), in Universidad Iberoamericana de Leon 0.16 tons are generated a day (Aguirre, 2008), in Universidad Autonoma Metropolitana 1.55 tons are generated a day (Espinosa, 2007; Espinoza et al., 2009), this generation depends on many factors, mainly on the number of people inside the institution, later on we will discuss the generation per capita for a more accurate conclusion of this information.

As we can see, not much information exists about the generation of waste inside Mexican universities (per capita, total waste generation, composition of waste, etc.) (Ojeda and Ramírez, 2007), although some universities have their Waste Management Program. Therefore, the main objective of this chapter is to show the results obtained from at least 3 years at the UTL, to establish as precedent and testimony to continue bettering the handling of waste inside educational institutions in Mexico.

5.2   BACKGROUND AND FIELD OF STUDY

The UTL is located in Leon, Guanajuato Mexico, and it was founded in 19953. In present time, it offers around 10 degrees for TSU (University Superior Technician, level 5B, a level before Engineering). Among these degrees, we have Environmental Technology. Besides this, an academic program exists in the afternoons to obtain the degree of Engineering. To offer the model 70–30 (70% practice and 30% theory), a lot of specialized labs exist to be used by the students; three terms exist per year that is January April, May August, and September December. Today, around 3,486 people are in the university, among which are students, teachers, and administrative personnel (see Table 5.1)

TABLE 5.1   Diverse sector population at Universidad Tecnologica de Leon in 2009 and 2010

	2009	2010
Students	1513	2834
Full time Professors	117	131
Administrative[1]	117	130
Half time Professors	274	391
Special projects	1	0
Service[2]	N.D	N.D

[1]Includes personnel in labs and information center.

[2]Includes Cafeteria service, cleaning and surveillance, which are external companies – constantly rotating, but only represent less than 2% of the population.

The Waste Management Program was started in 2007, with the storage center called Universitary Collecting Centre (UCC) or CUPA (Spanish acronym). Since then, the program offers service to the entire university by gathering, storing, and separating waste. Three categories exist for the waste generated in the university that is organic (green container), not organic (blue container), and garbage (black container). Not organic waste was sold for later recycling, the organic waste were used to produce natural fertilizer, and garbage is sent for final disposal at the sanitary landfill (see Figure 5.1).

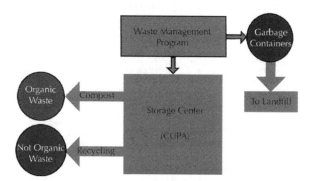

FIGURE 5.1   Operation diagram of the storage center and destination of the waste.

The whole university's community (students, professors, and administrative personnel) deposit waste in each containers, then the collecting is made (just organic and inorganic waste are taken to UCC).

5.3   MATERIALS AND METHODOLOGY

To satisfy the needs of the university in regards to the handling of waste, a diagnose was made through a quantification of the waste, with the purpose of determining the indicators of generation and designing a Waste Management Program according to the results obtained by the study. To complement the information obtained by the program, the following methodological steps took place:

5.3.1   Special Waste Handling Plan

The sources of waste generation were determined to make a qualitative analysis (separation indicators) and afterwards a quantitative analysis (as mentioned in point 2), through the establishment of each source, the kinds of waste were established from its generation to its sale; the specific needs and determining of responsibilities of certain aspects such as: the generation, containment, internal collection, storage, primary and secondary separation, the original plan written by Lopez in 2008 (López, 2008), and continuing with actualization through indicators of the section 3 (indicators calculation).

5.3.2   Quantitative Analysis

Three analysis were made: one from June 2 to June 6 of 2008, another from June 22 to July 18 of 2009, and the last from February 15 to March 6 of 2010, excluding Sundays; to quantify the production of waste from each source (the samples correspond to two terms from the university: January-April and May-August). Once collected, the samples were classified by source of generation, their physical properties were measured and determined (such as their density and volume), as also their composition.

5.3.3 Indicators Calculation

The indicators were determined from the sampling done in the quantitative description, to evaluate the efficiency of the plan, such as: total generation, *per capita* generation, percentage of waste recovered and kilograms of waste sent to the sanitary landfill, and kilograms sent to recycling. Also, the different factors that vary waste generation inside the campus were analyzed, some of which are: holidays, professional practice, graduations, cultural events, and others that are detailed further along.

Some of the data and indicators were obtained through the measuring of waste that entered the storage center (UCC), where the control of these measures is done through an electronic log:

- Waste subject to appreciation such as: PET, HDPE, cardboard, metal, aluminum, paper, and glass were collected, quantified and stored in UCC for later sale.
- With the organic waste, these were collected from the cafeteria, gardens and some of the waste containers in the university to elaborate natural fertilizer (compost).
- The wastes that do not have a recuperation potential or that cannot be used as natural fertilizer were deposited in garbage containers, where they are taken to the sanitary landfill. The measurements of this waste are used to create an operational performance indicator[1]: total of waste sent for final disposal (kilograms sent to the sanitary landfill).

5.3.4 Containment Infrastructure Proposal

According to the needs detected in each of the sources of generation as a result of the previous sampling, some proposals were made for the acquisition and distribution for the containment infrastructure for each of the buildings; including the capacity evaluation (m^3–cubic meters) of the general garbage containers (where they are stored until the local authorities take them to the landfill).

5.3.5 Environmental Education

An educational campaign was put into action for the entire university's community about the appropriate separation of the waste in the different containers inside the university with the intention of increasing the separation indicators and collecting of appreciable waste; and therefore, reducing the amount of waste that are sent to final disposal.

5.4   RESULT AND DISCUSSION

5.4.1   Special Waste Handling Plan

This handling plan includes, among other things: a qualitative analysis of waste and the different flow diagrams of the methods established for waste handling.

5.4.1.1   Qualitative Analysis

From the revision made, 16 sources of generation were identified inside the university, which are shown on Table 5.2.

TABLE 5.2   Waste generation sources related to specific activities

Source of Generation	Activities
Buildings (A, B, C, D, E, F)	Classes, administrative offices
Cafeteria	Catering
Laboratories (A, B, C)	Specialized education by degree
Link Center	Conferences, administrativework and publicity
Information Center	Book lending, magazines, etc.
Gardens	Fun and recreation
Football field and basketball courts	Fun and recreation
Maintenance	Facility and equipment maintenance
Construction areas and remodeling	Construction activities

Source: Modified and upgraded since López, (2008)
Note: all buildings have two floors, except the Information Center and Cafeteria.

TABLE 5.3   Waste classification in the Universidad Tecnologica de Leon

Inorganic		Garbage		Organic
Paper and newspapers	Books and notebooks	Metalized wrappings (cookies, potato chips)	Brochures	Cookies
Marker boxes	Aluminum	Paper wrappings	Plastic	Fruit
Folders	Invitations	Plastic sheets	Gum	Food scrap
Magazines	Paper	Plastic bags and tetra pack (juice, milk)	Compact Discs	Garden waste
Leaflet	Pieces of paper	Spoons		
Manuals	Soda bottles	Pens		
Calendars	Water bottles	Diapers		
Carton boxes	Yoghurt bottles	Toilet paper		
Pen boxes	Glass	Fruit containers		

Once identified the sources of waste generation, a qualitative description was made from each obtaining different results. Table 5.3 shows waste that is deposited in each of the containers according to their classification inside the university (organic, inorganic, and garbage). It is important to clarify that carton, office paper (books and notebooks), mixed paper (magazines and invitations), newspapers, and so, are considered as inorganic because each are subject to sale and recycling (that is the internal classification corresponding to the different containers).

Even though, people know in which container goes what kind of waste (due to environmental education), sometimes they do not deposit the garbage in its correct place. The results of a qualitative separation are shown below in Table 5.4 (this sample was taken in 2008).

5.4.1.2 Flows Diagrams of Waste Handling

Five flow diagrams were created according to the handling of each appreciable waste, which include: newspaper, office paper, mixed paper, organic waste, and carton, there also exist flow diagrams that include the handling of dangerous waste such as, electronic appliances, serigraphy waste, electronic devices, fluorescent lamps, used batteries, and waste from the different labs. Each procedure shows the specific needs and assignation of responsibilities for the handling of each waste. All the flow diagrams are available for the university's community through our quality website: http://calidad.utleon.edu.mx/access/index.php, and form part of the environmental aspects of SGA-UTL.

TABLE 5.4 Qualitative description of the waste containers

Source	Container		
	Organic	Inorganic	Garbage
Building A	S	S	R
Building B	R	I	I
Building C	S	I	I
Building D	I	S	S
Building E	N.D.	N.D.	N.D.
Building F	UN	R	UN
Lab A	S	S	R
Lab B	S	S	R
Lab C	S	I	I
Cafeteria	I	I	S
Link Center (CVD)	R	S	S
Information Center	S	I	I

UN = Unacceptable (separation between 0–25%).

I = Insufficient (separation between 26–49%).

R=Regular (separation between 50–75%).

S= Sufficient (separation between 76%–99%).

E = Excellent (separation to 100%).

N.D. = No data, because in 2008, the building was not in use, yet.

5.4.2 Quantitative Analysis

The results obtained from the sampling are shown in Table 5.5 (divided by sources of generation), including average weight, obtained in 2008, 2009, and 2010 with their respective standard deviation. In all the data, we can observe that the greatest generation of waste is produced by activities from the Cafeteria, and the least generation of waste is variable depending the year.

The volume and density measurements are shown in Table 5.6, where we can observe that the cafeteria is one of the highest in regards to volume and density, given that its composition is from organic waste (food) and Styrofoam.

TABLE 5.5 Average weight of waste (organic, inorganic, and garbage) for each source of generation

Year	2008		2009		2010	
Source	Weight (kg/Day)	Standard Deviation	Weight (kg/Day)	Standard Deviation	Weight (kg/Day)	Standard Deviation
Building A	12.89	5.76	12.70	12.76	9.0	7.3
Building B	19.22	6.86	20.44	10.86	23.94	14.85
Building C	13.60	13.47	9.77	5.83	37.57	18.00
Building D	25.67	9.15	2.86	2.77	5.83	1.05
Building E[1]	-	-	9.26	8.10	4.91	4.35
Building F	13.37	7.07	4.76	3.83	7.79	7.25
Cafeteria	64.91	11.04	25.01	14.87	47.24	24.41
CVD	12.27	11.87	4.28	2.17	6.31	7.51
Information Center	12.18	5.18	5.75	4.95	4.98	2.55
Lab. A	11.44	10.94	3.33	2.47	6.66	6.01
Lab. B	8.13	5.53	2.62	2.59	4.35	3.83
Lab. C	8.01	5.18	5.32	5.56	8.41	4.79
Paper containers[2]	-	-	17.67	12.19	3.09	3.05
Garden waste[3]	-	-	184.00	-	-	-

[1] In construction during 2008.
[2] In 2008 there aren't measurements of the paper containers.
[3] In 2009 the garden waste could just be sampled, the average of generation was obtained through one month of measuring.

TABLE 5.6 Average volume and density of waste (organic, inorganic, and garbage) by each generation source

Año	2008		2009		2010	
Source	Volume (m3)	Density (kg/m3)	Volume (m3)	Density (kg/m3)	Volume (m3)	Density (kg/m3)
Building A	0.22	57.31	0.66	56.71	0.24	38.39
Building B	0.50	40.97	1.81	45.11	0.33	71.48
Building C	0.23	57.28	0.70	48.22	0.71	53.29
Building D	0.38	67.83	0.33	64.51	0.15	38.20
Building E	-	-	0.62	44.11	0.09	52.26
Building F	0.11	123.50	0.48	29.25	0.17	44.95
Cafeteria	0.57	114.68	0.34	50.81	0.48	98.05
CVD	0.16	78.38	0.10	46.48	0.16	38.34
Information Center	0.16	74.90	0.09	92.44	0.14	36.84
Lab. A	0.16	69.51	0.19	42.17	0.17	39.12
Lab. B	0.16	52.36	0.08	38.42	0.07	59.21
Lab. C	0.14	55.45	0.13	35.99	0.19	44.71
Paper containers.	-	-	0.20	72.91	-	-
Garden waste.	-	-	2.35	313. 64	-	-

In Table 5.7, it can be observed the composition of the waste generated inside the university, having as the greatest generation of waste the organic matter (composed mainly by the cafeteria waste) and toilet paper, in this composition are excluded the garden waste due to that its generation is seasonal (spring, summer, autumn, and winter).

5.4.3   Indicators Calculation

The total average generation of waste in 2008 was 202.065 kg/day, with a standard deviation of 92.055 and a *per capita* generation of 0.08 kg/person a day, with a population of approximately 2,525 people. In that year, the storage center recuperated the 25.6% of the total waste generated in the university, which represents 48% of the recoverable waste (as shown in Figure 5.2).

In 2009, the total average generation of waste was 147.47 kg/day and the *per capita* generation was 0.05 kg/person a day, and the percentage of recovered waste in the storage center was 29.76%, it is important to mention that the goal for the SGA-UTL was 25%, so from that moment on we started improving the environmental education program for the university community, although this number was good, the percentage of appreciable waste decreased from 46% in 2008 to 34.1% in 2009.

Finally, in 2010 the total average of generation went from 220.01 kg/day and a *per capita* generation 0.063 kg/person a day, and yet we cannot compare with 2008 and 2009 given that 2010 is still in progress while this chapter is being written and we do not have the results for all the year.

During work days in the university in 2008, 2009, and 2010 (present) (Turcott et al., 2008; Turcott et al., 2009), the measurements were made of the appreciable waste that entered the storage center. Figure 5.3 shows the amount of waste that entered the storage center for appreciation, since January until December (Juarez et al.,2008, 2009, 2010).

TABLE 5.7   Percentage by weight

Waste	2008	2009	2010
Organic matter	33.97%	23.45%	29.80%
Toilet paper	21.35%	18.00%	15.38%
Carton	8.18%	5.33%	4.10%
PET	8.10%	5.46%	5.00%
LDPE	3.12%	3.70%	4.00%
Markers	2.89%	-	0.03%
Glass	2.88%	2.77%	9.00%
File paper	2.59%	4.52%	3.04%
Styrofoam	2.08%	2.03%	4.00%
HDPE	2.04%	1.89%	2.00%
Polypropylene	1.98%	1.24%	6.00%
Tetrapack	1.93%	1.71%	1.69%
Waxed paper	1.73%	1.07%	-
Napkins	1.32%	6.77%	-
Mixed paper	1.07%	4.96%	1.50%
Aluminum	1.02%	0.82%	1.40%
Waxed carton	0.84%	0.35%	-
Newspaper	0.80%	0.43%	0.13%
Metal	0.64%	0.96%	1.00%
Garden waste	0.51%	0.39%	-
Construction waste	-	6.14%	-
Electronic waste	-	3.39%	-
Wrappings	-	1.40%	3.00%

TABLE 5.7   *(Continued)*

Waste	2008	2009	2010
Oil	-	-	1.00%
PVC	-	-	1.00%
Fine waste	0.73%	3.11%	6.12%
Other*	0.23%	0.12%	0.81%

* Includes: sponge, dust, mop, clothes, cotton, gauze, batteries, porcelain, CDs, soap, cord

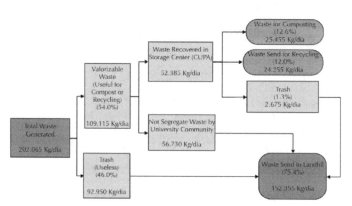

FIGURE 5.2   Balance generation, recuperation, and no recuperation of waste in 2008.

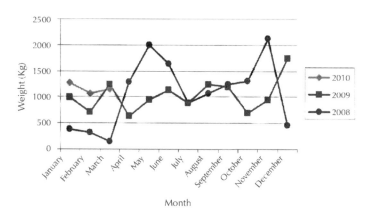

FIGURE 5.3   Amount of recoverable waste that entered the storage center for their recuperation in 2008, 2009, and partially in 2010.

In 2008, the total amount of waste recovered in the storage center was 14.52 tons, in 2009 it was 12.33 tones, and data for 2010 is still unavailable. It is important to clarify that even though the recovering of waste was greater in 2008 than 2009, the percentage of recovered waste was greater in 2009 as it was mentioned before.

The composition of the waste separated for their appreciation and later sale of conversion into natural fertilizer, is shown on Table 5.8. In the column of 2008 is included some waste of 2007 (November and December).

From the cafeteria, the greatest amount of recovered waste is organic, with the only objective of producing natural fertilizers monthly, to be used in the green areas of the university. Also, a great deal of carton was generated and office paper due to its consumption in office areas and this is sold for recycling.

Approximately 95% of the waste that enter the storage center are recovered, on an average 5.1% of the waste are returned in the general garbage containers, which will be taken to the sanitary landfill (see Figure 5.2). According to the chart, the waste that is generated in least amount include: wood, aluminum, and ferrous metal. In this case, aluminum is collected by the cleaning personnel before it arrives to the storage center.

The generation inside the university is affected by diverse factors, shown in Table 5.9. One of the main factors that influences in the increment or decrement of the waste is the amount of students on each term. One of the main factors for decreasing waste production is holidays.

TABLE 5.8   Composition of recovered waste in the storage center

Waste	Kg Recovered 2008	Kg Recovered 2009	Treatment*
Aluminum	54.27	76.42	Recycled
Carton	1215.83	393.3	Recycled
Wood	10.20	-	Compost
Ferrous Metal	42.31	71.62	Recycled
Organic waste	6907.23	7040.4	Compost
File paper	1948.44	2316.3	Recycled
Mixed paper	630.26	511.6	Recycled
HDPE	182.65	374.5	Recycled
Newspaper	1084.610	279.4	Recycled
PET	626.544	754.55	Recycled
Glass	354.790	507.55	Recycled
Trash from the storage center	738.296	N.D.	Landfill

*Recycling and the landfill are outside the university.

5.4.4   Containment Infrastructure Proposal

Figure 5.4 shows the external islands that are used for the separation of waste; each island has three containers: the blue one is used to collect inorganic waste, the green one for organic waste, and the black one for garbage. The university has 27 islands for separating and containing the waste outside the buildings; the capacity for each island is 0.488 m³, having a maximum capacity of 13.16 m³. There also exist three general containers for garbage, where waste is contained to be taken to the sanitary landfill, where two containers have the capacity of 7.46 m³ and the other one 9.5 m³.

In the same way, 33 islands exist inside the buildings to separate the waste, these islands have less capacity (each one has a capacity of 0.233 m³), for a total capacity of 7.7 m³. There also exist 24 containers to separate paper and carton. The location of each container is specified by the sources of generation explained before.

For the operation of the storage center and separation of waste, we have a space of 147 m² (square meters) with walls and a roof, and we also have a New Holland vehicle that is used for the internal collection of waste and maintenance of the natural fertilizer.

According to the previous analysis, a requisition of more external and internal islands was made to cover the needs of the new areas, as also the acquisition of more paper containers. In the case of the waste containers a requisition was made for an additional container to satisfy the needs for an adequate contention capacity. The average density of the waste was 43.13 kg/m³ so a new waste container was required to avoid cleanness issues and to have the capacity required for contention special events take place at the university.

TABLE 5.9   Monthly factors involved in the generation y recuperation of waste in 2008

Month	Factor(s)
January	The collecting of waste was affected by the little knowledge from the university's community about the program of waste handling.
February	The generation decreased because of the amount of holidays in the university.
March	The decrease of the generation was affected because of holidays (two weeks: holy week).
April	The increase of waste was caused by a special event called "Jornadas". Usually different kinds of cultural events happen: sports, workshops, conferences, get-togethers. In these events the amount of certain waste increases (organic, carton).
May	The increase of waste (generally organic, office paper and mixed paper) is due to the meals given to teachers during staff training week; which occurs the last week of the term. From this month till the end of summer, the generation of garden waste increases.
June	The high recovering of appreciable waste was because a description was made this month, additionally we had the celebration of TSU day (student's day) that generates a lot of organic waste, carton, PET, and food leftovers.
July	Waste generation was affected by summer vacation (2 weeks)
August	Once more the generation and recovering of waste was affected by vacation time at the end of the term. Organic waste increased because of the staff training week to professors.
September	Because the new students who enter the campus don't know the waste handling program it affects the recovering of appreciable waste and increases garbage generation.
October	Educating the entire community about the correct way of separating waste caused an increment in the recovery of appreciable waste.
November	Waste increased because of graduation ceremonies and other events inside the university.
December	The decrease of waste was due to vacation time in this month.

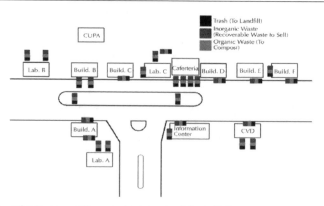

FIGURE 5.4   Distribution of the exterior islands of the buildings.

5.5 ENVIRONMENTAL EDUCATION

An environmental education campaign was put into action to demonstrate to the university community about the correct separation of the waste in the different containers inside the university (awareness campaigns, video, surveys, talks, forums, etc.) with the purpose of incrementing the indicators of separation and recovering of appreciable waste, and reducing the amount of waste sent for final disposal. The administrative personnel, teachers, and students that were trained during this campaign are shown on Table 5.10.

TABLE 5.10 Number of people trained by area

Program or Area	Number of People Trained 2008	Number of People Trained 2009
Industrial Electromechanics	390	351
Information Technologies	484	361
Economical Administrative	563	551
Sustainability for Development1	155	95
Guided Visits	99	0
Administrative and supporting personnel	60	7
Others2	-	640
TOTAL	1751	2005

[1] Includes the degree of Environmental Technology.
[2] Includes students from Engineering and reinforcement in environmental education to students of 2008.

5.6 CONCLUSION

The generation of waste inside the university is variable due to various factors, such as: number of students, holidays, and special events, among others. The UTL is not alien to these factors. The Waste Management Program has worked since 2008 in conjunction with the storage center with the intention of following up to all actions focused towards a comprehensive handle of waste. To increase the efficiency of the primary separation, permanent environmental educational campaignshave been implemented for teachers, administrative personnel, directors, cleaning personnel, and students. All this has the intention of creating awareness of the importance of waste handling inside the university. The recovering percentage in 2009 it was better than 2008, and in the beginning of 2010 the amount of waste recovered seems to be better than 2008 and 2009. In fact, January 2008 is not representative due to the fact that the program was barely starting, and in that year a lot of factors affected the generation of waste, but by 2009 and 2010, we expect that the consolidation of this program could be clearly perceived.

The generation *per capita* calculated during the sampling was 0.08 kg/day (2008), 0.05 kg/day (2009), and 0.063 kg/day (2010), compared to other universities in Mexico, it is inside a range of +0.02 to −0.05 kg/person a day:

- Universidad Iberoamericana de León, Guanajuato, México: 0.041 kg/person/day (Reyes, 2008).
- Universidad Autónoma Metropolitana, México: 0.110 kg/person/day (Espinoza et al., 2009; Espinosa et al., 2007).

- Universidad Autónoma del Estado de Morelos 0.082 kg/person/6 day approximately (Ortiz et al., 2007).
- Tecnológico de Monterrey campus León, Guanajuato, México 0.0963 kg/person/day (Turcott et al., 2010).

Nevertheless, it can be observed that it is a low indicator compared with the generation *per capita* in Mexico for 2008, which was 0.97 kg/person/day[2], the UTL only generated in 2008 and 2009: 86.33 tons of waste. From which nearly 26% were recovered for their later appreciation. The organic wastes are recovered in great quantities and were used to produce natural fertilizer. This is why the Waste Management Program is the main component of SGA-UTL. Of the universities in the United States, Brown University recycles 31% of the waste, the University of Florida 30%, and finally one of the most successful programs of waste handling is the University Santa Clara in California, which recycles around 50% (Reyes, 2008).

The composition compared to other universities, for example, Universidad Autonoma de Baja California Mexicali I campus, the waste with greater generation in buildings was paper, and in their gardens and their community center were organic waste (between 54% and 80%) (Reyes, 2008), in the cafeteria of the UTL (between 24 and 34%) and in its gardens is where the most organic waste are generated.

The special waste handling plan, which is the base for all the operation of the storage center, exists since 2008. Nevertheless, all the indicators inside the continuous betterment cycle, shown and discussed here, are in constant upgrading.

Even though there were a lot of activities in 2008 and 2009 in regards to the Waste Management Program, there is still room for betterments, from facilities of the storage center to betterments in the process of collecting waste and measuring indicators. To achieve these betterments, it is required a multi-task job in conjunction with other areas of the university, because the handling of solid waste has a lot of complex components that require different abilities and knowledge to find the best solution to this problem. Since 2009, we have looked for the cooperation of other degrees, as well for technological development.

All that has been mentioned here has required a significant investment in time, money, and effort from teachers, students, and personnel in general. Nevertheless, it is the duty of each educational institution to generate knowledge and to generate innovation in technology to solve environmental problems, and the most important to teach by example. Also with actions that will lead us to be coherent with what is taught inside the classrooms, especially for students of environmental degrees.

For 2010, Rectory has expressed their interest in obtaining an ISO 14001 certification, which will support and motivate a lot of the activities mentioned before, but this also implies a great challenge in terms of time, effort, and investment.

Therefore, we can conclude that a lot of work is still needed and a lot of future challenges will have to be overcome to achieve significant advances in waste handling, especially in Mexican universities, because depending on the advances obtained, this will help to minimize the negative impact caused by the same universities. This chapter is a contribution to demonstrate specific and detailed indicators, and real life experience by implementing a Waste Management Program, because few universities document their achievements and contribute with real changes in Mexico for the institutions that are starting their own programs and for the rest that need to better their established programs.

ACKNOWLEDGMENT

We want to thank the UTL for providing the financial resources to operate the Waste Management Program, as also to all the university's community for participating in the different activities.

We also want to thank CONCYTEG (Consejo de Ciencia y Tecnologia del Estado de Guanajuato) for their financial contributions to the realization of the quantitative analysis of 2009.

We want to thank Jose de Jesus Mendoza Rivas and Julian Barragan Diaz, English coordinators of the UTL, for revising the English version of the essay and for all their comments to improve the chapter, and thanks to the Sergio Rico for the final translation and review of the chapter.

To all the professors of the Environmental Technology degree, especially Javier Paramo Vargas, who started a lot of activities of the Waste Management Program.

To Benito Juarez Rodriguez who was the first operator of the storage center and for his participation in a lot of the activities of UCC during 2008.

To the students of the Environmental Technology degree: Beatriz Padilla Rizo, Jose Salud Lara Servin, and Israel Rico Vera for their invaluable participation in the waste sampling of 2009.

And last but not least, a special acknowledgment for all the students of the Environmental Technology degree that have participated since 1999 till 2010 to maintain alive this project.

KEYWORDS

- **Composition of the waste**
- **University campus**
- **Valued waste**
- **Waste management**

NOTES

[1] NMX-SAA-14031-IMNC-2002. "Gestión Ambiental-Evaluación del desempeño ambiental-directriz". Instituto Mexicano de Normalización y certificación, A.C., México (2003). Translation of the norm ISO 14031 (1999).

[2] Secretaría de Medio Ambiente y Recursos Naturales, "Generación total y per cápita de residuos sólidos urbanos". National system of environmental and natural resources. Basis indicators set, urban solid waste; Pressure indicator 4-2 (2008), Available from: http://www.semarnat.gob.mx/informacionambiental/Pages/index-sniarn.aspx, [Accessed: July 1st, 2009].

[3] Universidad Tecnologica de León. Quienes somos?: Historia de la UTL", UTL's oficial Web page., [Online]. Available from: http://www.utleon.edu.mx//index.php?option=com_content&task=view&id=22 &Itemid=95. [Accessed: March 26, 2009].

REFERENCES

Aguirre, M. *Plan de manejo de residuos para la Universidad Iberoamericana de León*, Superior Universitary Technician disertation, Universidad Tecnologica de Leon, Leon, Gto., México (2008).

Espinosa, R. M., Turpin, S., Polanco, G., De la Torre, A., Delfín, I., and Raygoza, M. I. *Programa de gestión integral de residuos sólidos en la UAM Azcapotzalco: Una experiencia camino al éxito*, First meeting of solid waste experts, Universidad Autónoma de Baja California, México, Baja California, pp. 336–347 (2007).

Espinoza, R. M., Turpin, M. S., A., de la Torre, Vázquez, R. C., Delfín, I., González, B., and Cisneros, A. L. *Proceso de implementación de un sistema internacional de administración de calidad en un programa de gestión integral de residuos sólidos en la Universidad Autónoma Metropolitana-Azcapotzalco*, Second meeting of solid waste experts, Universidad Michoacana de San Nicolás de Hidalgo, México, Michoacán (2009).

Estrada, B. I., González, J. M., Hernández, F. G., Pérez, V., Piña, G. A., and Regalado, M. P. *Asesoría en el Sistema de Gestión Ambiental en la UTL*, Superior University Technician disertation, Universidad Tecnologica de Leon, Leon, Gto., México (2007).

Juarez, B., López, K. G., and Reyes, M. B. Bitácora electrónica de entradas de residuos de manejo especial y peligrosos. *Electronic Book in Microsoft © Excel*, from UCC. Universidad Tecnologica de Leon, México (2008, 2009 and 2010).

López, K. G. Plan de manejo de residuos para la Universidad Tecnológica de León, *Superior Universitary Technician dissertation*. Universidad Tecnologica de Leon, Leon, Gto., México (2008).

Ojeda, C. S. and Ramírez, M. E. *Caracterización de residuos sólidos: el potencial de reciclaje para una institución de educación superior*. First meeting of solid waste experts, Universidad Autónoma de Baja California, México, Baja California, pp. 76–89 (2007).

Ortiz, M. L., Sanchez, E., and Lara, J. *La gestion mediambiental en la Universidad Autonoma del Estado de Morelos*, Mexico. IV Internacional seminary of University and environment, Gestion ambiental institucional y ordenamiento de los campus universitarios compiled by O. Saenz, Bogota, Colombia, pp. 77–90 (2007).

Palacios, J. L. Un caso de estudio: Universidades Verdes. CONCYTEG's *Gazette ideas*, Year 3. Number 32. CONCYTEG, México, pp. 41–43 (2008).

Reyes, M. B. *Implantación de controles operacionales para el SGA-UTL*, Superior University Technician disertation, Universidad Tecnologica de Leon, Leon, Gto., México (2008).

Turcott, D. E., López, K. G., and Reyes, M. B. *Informe anual del programa de manejo de residuos*. Universidad Tecnologica de Leon, México (2008).

Turcott, D. E., Muñoz, L., López, K. G., and Reyes, M. B. *Informe anual del programa de manejo de residuos*. Universidad Tecnologica de Leon, México (2009).

Turcott, D. E., Muñoz, L., López, K. G., Reyes, M. B., Murillo, M. B., and Chávez, V. *El manejo de residuos en las instituciones de educación superior: El caso de las Universidades en Guanajuato*. Third meeting of solid waste experts. Mexico, Distrito Federal, Universidad Autonoma Metropolitana, vol. 1, pp. 9–11 (June 2010).

6 Solid Waste Management by Application of the WAMED Model

Viatcheslav Moutavtchi, Jan Stenis, William Hogland, and Antonina Shepeleva

CONTENTS

6.1   INTRODUCTION

This chapter aims to develop a general model for the evaluation of ecological–economic efficiency that will serve as an information support tool for decision making at the corporate, municipal, and regional levels. It encompasses cost-benefit analysis (CBA) in solid waste management by applying a sustainability promoting approach that is explicitly related to monetary measures. A waste managements efficient decision (WAMED) model based on CBA is proposed and developed to evaluate the ecological–economic efficiency of solid waste management schemes. The employment of common business administration methodology tools is featured. A classification of competing waste management models is introduced to facilitate evaluation of the relevance of the previously introduced WAMED model. Suggestions are made for how to combine the previously introduced EUROPE model, based on the equality principle, with the WAMED model to create economic incentives to reduce solid waste management-related emissions. A fictive case study presents the practical application of the proposed CBA based theory to the landfilling concept. It is concluded that the presented methodology reflects an integrated approach to decreasing negative impacts on the environment and on the health of the population, while increasing economic benefits through the implementation of solid waste management projects.

During the past few decades, there has been an increase in the amount of scientific work concerning the evaluation of efficiency, in ecological, economic, and social terms, of solid waste management (SWM) to assist decision-making support systems (Björklund, 2000). This increase is due to:

- The need to solve the SWM problem in the context of ensuring the vital activity and environmental safety of regions by simultaneously decreasing its negative impacts on the environment and population health, while increasing the economic benefits through the implementation of SWM projects
- Substantial changes in SWM, including the compulsory source separation of waste, the development and implementation of technologies based on use of secondary materials, technical modernization of waste treatment facilities, that have been initiated by changes in the legislative–normative base (Council of the European Communities, 1991; Council of the European Union, 1999) and by changes in the chemical and morphological nature of municipal solid waste (MSW)
- Increasing competition and the transition of the process of collection and recycling of MSW in developed countries into a profitable business
- The increasingly relevant fact that it is not possible to solve the SWM problem by means of a single environmental protection measure owing to the complex nature of the waste generated (European Parliament and the Council of the European Union, 2000).

This work is based on a general overview of existing SWM models and methods. The specific aim is to develop a model for the evaluation of the ecological–economic efficiency (ECO-EE), defined as "the ratio of man-made capital services gained to natural capital services lost as a result," (Groom et al., 2005) that will serve as an information support tool for decision making by SWM actors at the corporate, municipal, and regional level when, for example, deciding on introducing a new system or modifying an existing one.

A literature search showed that earlier studies within the current field have dealt with, for example, waste management models and their application to sustainable waste management. Thereby, the models currently being used within SWM have been reviewed and their major shortcomings have been highlighted, implementing a division into those based on (i) cost-benefit analysis (CBA), (ii) life-cycle assessment (LCA), and (iii) multi-criteria decision making. Also, the following review serves as a background for the later evaluation of gaps, in a scientific sense, filled by the waste managements' efficient decision (WAMED) model. The present study is a general extension of an earlier attempt by the authors to apply the WAMED model to MSW in particular. (Moutavtchi et al., 2008)

The different methods used for supporting waste management decisions can be useful on a number of different levels in society and can be described as system analysis tools. Although, the expectations of system analysis tools often are quite high, the expectations are sometimes not met. One reason may be that the wrong method was chosen. Another reason may be that the data and methodological uncertainties are so large that clear conclusions are difficult to draw. (Finnveden et al., 2006)

However, no single model is found to consider the complete waste management cycle from the prevention of waste to final disposal, or to be fully sustainable, or to consider the involvement of all relevant stakeholders. No model examined considers environmental, economic, and social aspects together in the application of the model. (Morrissey and Browne, 2004) Nevertheless, economic and social CBA is the form of economic analysis performed by most environmental protection agencies today. (Vigsø, 2004)

However, earlier models developed by Roberge and Baetz (1994) were meant to be used within a long-term industrial context with emphasis on waste reduction, where consideration was planned "in the future" to be given also to environmental, political, social, and legislative factors, as well as to the limitations on pursuing an economic analysis with technical and policy considerations. Their new model is claimed to allow for consideration of the economic realities and budgetary constraints of industry.

Thereby, the problem is formulated as a general mixed integer linear programming (MILP) problem involving the study of the main existing waste streams with the related waste reduction connected to the current waste reduction projects. This approach does not encompass the CBA approach and it does not explicitly consider sustainability aspects. And also it is not the case for earlier MILP approaches, for example, by Gottinger (1991), that, as is usually the case in this context, emphasize mathematical modeling algorithms for mainly waste management route generation objectives.

Attempts by, for example, Kijak and Moy (2004), do have the goal of achieving sustainable waste management practices, in this case by balancing global and regional environmental impacts, social impacts at the local community level, and economic impacts. In doing so, spatial resolution is introduced into the LCA process to account for the impacts by use of a regional scaling procedure for LCA data for emissions to the environment. The pressure–state–response (PSR) model, suggested by the Organization for Economic Cooperation and Development (1993), is applied to introduce a relationship between the model boxes representing pressure, state, and response and to identify the flows of material and burdens between these model boxes. Spatial resolution accounts for site-specific assessment of environmental emissions with regional consequences. The model so far represents a loose general framework later to be transformed into Excel spreadsheets that automate the calculations and modeling. This model approach uses the conventional full cost accounting (FCA) methodology, with some modifications, for example, to identify the economic impacts. However, this approach does not explicitly encompass the CBA approach for example, employing the most commonly used business administration tools, even though it considers sustainability aspects.

More recent models for evaluating the overall resource consumption and environmental impacts of SWM systems (by Kirkeby et al., 2006 for example) use LCA to identify the most environmentally sustainable solution. Thereby, the impacts are combined with normalizations and weightings; it is possible to include economic costs in the model at a later stage, and odor, dust, noise, ethical issues, and social willingness towards a scheme are omitted but are important to consider.

Thus, the general version of the WAMED model introduced here is regarded as fulfilling an urgent need to provide a practically useful methodology to cover substantial parts of the SWM cycle and to provide a basis for decision making expressed in monetary terms, for example. In doing so, CBA is applied and the present value of benefits of the project or the policy being evaluated preferably exceeds the present value of the related costs. Generally, but not universally, economists tend to favor CBA as the tool for choosing projects. (Pearce et al., 2006)

Academic work has so far regarded sustainable development to be an aggregate or macroeconomic goal. Little attention has been paid to the implications of notions of sustainability for CBA. Only a handful of recommendations exist regarding the way in which CBA appraisals can be extended to take account of recent concerns of sustainable development (Kirkeby et al., 2006) in monetary terms.

The CBA is a methodology that, it is claimed, has the ability to handle a wide range of problems. These include the ability to capture important aspects of problems and to judge how far a policy or a project moves society towards some socially defined and accepted goal. Hence, CBA can be applied to any decision that involves a relocation of resources within society (Hanley, 1999). The CBA shows distinctive characteristics necessary and sufficient for use in decision-making support systems in SWM management (European Commission, 1996). The ability of different waste management models to accommodate CBA has been analyzed by the authors and is shown in Table 6.1, even though no comparative analysis of the qualities of the listed models is presented in the table itself.

As shown in Table 6.1, all waste management models presented include, to a certain extent, cost calculations for implementation of SWM scheme. Regarding the accounting procedures for the SWM schemes, the authors mainly concentrate on the economic results, which are subject to direct monetary estimation (profits made from selling secondary resources, energy, and compost), and environmental impact in the form of emissions.

TABLE 6.1   Ability of different waste management models to accommodate the procedures of cost–benefit analy sis (CBA)

Model	Calculation of costs incurred when implementing a scheme	Estimation of financial benefits when implementing a scheme	Accounting of environmental effects when implementing a scheme	Collation of costs and benefits
WastePlan[18]	Full cost accounting: WastePlan facilitates the use of full-cost accounting (FCA), an approach aimed at accounting for and allocating all the cost for solid waste management to appropriate programs (i.e., recycling, composting, collection, disposal) and management categories	Economic benefits: • Avoided cost disposal; • Source reduction: avoided cost of finished goods; • Recycling: value of recycled commodities Energy benefits: • Avoided use of energy in material extraction, production, and disposal processes	Air and water pollution benefits: • Avoided emissions in material extraction, production, and disposal processes (reduced mass emitted) Land use benefits: • Landfill space preserved through source reduction and recycling (volume, area); • Avoided resource extraction (forest area)	CBA, least-cost system planning, capacity analysis/ system mass balance assessment, sensitivity/ scenario analysis

TABLE 6.1 *(Continued)*

Model	Calculation of costs incurred when implementing a scheme	Estimation of financial benefits when implementing a scheme	Accounting of environmental effects when implementing a scheme	Collation of costs and benefits
IWM-2[19]	Inputs: • operating costs • energy requirements	Outputs: • energy • recovered materials • compost Avoided burdens from recovered materials and energy	Outputs: • air emissions • water emissions • residual solid waste (landfill volume)	The inputs and outputs are all done on a mass basis
WISARD5[20]	The calculation of costs is presented for each type of collection system in terms of capital expenditure and financing, operating expenditure, site management, administration, monitoring, closure and aftercare, and insurance among others	Revenue from energy, compost, and recycling	Air emissions, water emissions, and emissions to soil	CBA, Mass balance
EPIC/CSR[21]	Collection, processing, and administration costs (the tipping fee charged at facility, actual capital costs of equipment, and infrastructure and operating costs). The Model provides default values for processing costs associated with recycling, composting, energy-from-waste facilities and landfill facilities	Revenue: • energy-from-waste program • recycling program • composting program	Environmental impacts: • energy consumption • greenhouse gas emissions (climate change) • emissions of acid gases (acid precipitation) • emissions of smog precursors (smog formation) • air emissions of lead, cadmium, mercury, and trace organics (health risk) • water emissions of heavy metals, dioxins, and biological oxygen demand (impact on water quality) • residual solid waste (land use disruption)	The environmental impact is determined by the model's life-cycle inventory module. The economic implications are ascertained by an economic analysis module. These modules can be used together or independently
MSW-DST[22,23]	Typical capital and operating costs for residential, commercial, institutional, and industrial sectors	Revenues generated through the sale of recovered materials (recyclable revenues), compost, fuels, (gas) energy	Environmental emissions (air, water), energy demands, landfilling of ashes	Balancing the cost and environmental aspects to provide a win–win solution. Minimum-cost strategy. The most cost-effective strategy
EUGENE[24]	The annual collection and transportation costs, the annualoperating and maintenance costs, the investment costs, the importation costs (from external sources), the salvage values of the technologies	The annual revenues from sales to the markets	The environmental and spatial indicators (will be integrated)	Total discounted net system cost
ORWARE[1,25]	Net costs include costs for investment and operation, spreading of residuals, and gas utilization. Costs for compensatory production of functional units are included as well	Recovered energy is valued at market prices for compensatory generation of heat and power	Emissions into air, water, and soil. Calculation of degradation products, energy output, primary energy carriers, heavy metals, nutrients	Life-cycle cost analysis

TABLE 6.1   *(Continued)*

Model	Calculation of costs incurred when implementing a scheme	Estimation of financial benefits when implementing a scheme	Accounting of environmental effects when implementing a scheme	Collation of costs and benefits
EASE-WASTE[14]	Impact characterization, normalization, and weighting	Revenue from remanufacturing, reuse of land, and reuse in construction	Emissions into air, water, and soil	Life-cycle cost analysis
MARKAL[18]	The investment costs (which are proportional to the installed capacity), fixed annual costs (proportional to the installed capacity), variable costs (proportional to production volume), delivery costs	Energy recovery, waste recycling	Greenhouse gas emissions, resource use, land use, waste volume	The identification of least-cost system configuration, the evaluation of the effects of prices. The identification of cost-effective responses to restrictions on emissions
MWS[26,27]	Total annualized cost for the national waste management system	Revenues: • recovered materials • compost • recovered energy (heat)	The environmental assessment: • the accounting of emissions to air and residual content of harmful substances in the waste or recovered material • the introduction of emission constraints and fees	The effect of different levels of cost increases, revenues for energy (heat), for compost, and for recovered materials

Thus, analysis of the available waste management models using CBA has shown that, at present, assessment of the monetary damage to the environment during implementation of a SWM scheme (showing the current damage) or of the possible positive economic results accruing from a change in the scheme (showing the prevented damage) is not offered by these models in an explicit form. Generally, even though waste treatment economics and management seem to have gained momentum over the past 10–15 years, publications on cost figures are scarce. This may be because waste treatment facilities are quite costly and only limited information is available for getting good cost estimates. Moreover, the literature is poor and only scattered data are available. Finally, the estimates based on statistical data corresponding to facilities built in the past have obvious shortcomings. Therefore, the present study can be said to represent an innovative attempt to encompass CBA into SWM in a more reliable way by applying a sustainability promoting approach that is explicitly related to monetary measures.

6.2 METHODOLOGY

In this chapter, a cost structure is proposed for evaluating the ECO-EE of SWM schemes. As a background, the introduction reviews, the reasons for the recent increase of related scientific research, the relevant earlier studies, and the current scientific frontline within the field of interest are presented and commented upon to give the scientific context. This is followed by an attempt to classify the available waste management models as a basis for the evaluation of the relevance and novelty of the introduced WAMED model that follows. A brief overview of the practical aspects concerning the application of CBA in primarily SWM is given. It provides a basis related to economic model methodology for the following exploration of the possibilities to set up a theory for cost and benefit structures for evaluation of the ECO-EE of SWM schemes employing the WAMED model and the company statistical business tool for environmental recovery indicator (COSTBUSTER) models. These models are combined with the EUROPE model based on the *equality principle*. Thereafter, the determination of benefits of ECO-EE during evaluation is analyzed, this being the most difficult part of the practical application of CBA and serving as the theoretical evaluation basis for the case study that follows. Thereby, the process of waste managements' decision making is summarized graphically to facilitate decision making. The case study concerns the cost-based evaluation approach exemplified by representative, modified data from a fictive landfill to demonstrate the general applicability of the introduced model. In the results and discussion section, important theoretical aspects and the practical application of CBA in SWM are discussed as the basis of the WAMED model. In a conclusions and recommendations section are summarized the features of the WAMED model, practical aspects of CBA related to the landfilling concept, implications for carrying out selection of scheme variants, and general recommendations for the application of the WAMED model and the COSTBUSTER indicator model. Generally, the short-term perspective (up to 5 years) is emphasized due to the overruling ambition to provide a tool for facilitating decision making as regards whether to invest in a SWM scheme or not. When the long-term effects become noticeable, it is usually time to cash in and sell, preferably with the profit that the investor originally hoped for. The validity and the reliability of the study is analyzed and future directions are given as suggestions for further research.

6.3   CLASSIFICATION OF WASTE MANAGEMENT MODELS

Computerized waste management models are typical examples of modern applied research work that can be used to support integrated ECO-EE-related administrative decision making in the SWM sphere. The objectivity of these and other models depends on their adequacy with respect to: (i) the aims and tasks of the practice of management in reality, (ii) the selection criteria of the most efficient investment projects, (iii) aspects of the development of environmental policies, (iv) purposeful programs and business plans, and (v) the stages of substantiation of certain measures in SWM. In Figure 6.1, a classification of waste management models developed by the authors is proposed using these five criteria of model objectivity and information concerning the decision-making modeling. Specifically, the criteria in Figure 6.1 refer to the use of the information concerning the modeling to facilitate decision making. Note that the different criteria may be used for evaluation of different models. For the sake of clarity, in Figure 6.1 the best correlated kinds of models are presented next to their primary criterion for evaluation according to the authors' judgments.

The models that are classified are selected among similar models that are regularly discussed in the scientific literature and, according to the authors, have characteristic features needed to identify the models. The selection is based on the relevant use of information for decision making in waste management, viewed from an administrative perspective.

The selected models were used for development of the classification based on the availability of features that support decision making as regards SWM solutions. Based on the usage of the results of the modeling in terms of the time, scope, and scale of decisions, it has been possible to position the WAMED model among tools aimed at facilitating the decision making process to provide economic and environmental results of different technological options and to show the model application framework from a managerial point of view. Also, certain models, such as IWM-2, have been used for the development of the methodological basis of the WAMED model as regards the model's scope.

FIGURE 6.1   Classification of waste management models (by the authors). MSW, municipal solid waste. (1), for example IWEM, industrial waste management evaluation model; (2), for example IWM-219, integrated solid waste management; (3), for example MIMES/waste model, a model for integrating the material flow with the energy system; (4), for example the MWS Model; (5), for example Chang et al. (6), for example Barlishen and Baetz, Everett and Modak (7), for example ORWARE 1, organic waste research; (8), for example EUGENE, optimization based decision support system for long-term integrated regional solid waste management planning (translated from the french language); (9), for example ETH Model, environmental evaluation of waste treatment processes with the help of life-cycle assessment; (10), for example MSW-DS, municipal solid waste-desision support tool; (11), for example EPIC/CSR Model, environment and plastics industry council/corporations supporting recycling; (12), for example FMS Model, environmental strategies research-FMS; (13), for example EASEWASTE (14), environmental assessment of solid waste systems and technologies.

The selected models are mentioned in the legend of Figure 6.1 but not in the figure itself. Thereby, the reader obtains the general picture of the proposed classification.

Also, this enables an overview of the nature of the SWM models used as basis for the development of the WAMED model.

The selected models are well known within the field of SWM. From a national perspective, certain models are more recognized and, therefore, better known in particular countries, for example, ORWARE in Sweden and MSW-DST in the USA.

In parallel, models such as IWM-2 have received international recognition, being well known along with nationally used SWM models. Useful results are produced as a result of, for example, international workshops (Chang et al., 1996a; Gielen, 1998; Ljunggren, 1998; Sundberg, 1998), where well known SWM models, national as well as others, are cross referenced in certain chapters.

6.4 THE WAMED MODEL

The WAMED model, developed by the authors of this work for evaluation of the ECO-EE of a SWM scheme, is, according to the proposed classification:

- A single-purpose, complex, short-period model for general use at the corporate, municipal, and regional level.
- Based on the structure for economic analysis outlined in by Moutavtchi et al.
- A model that considers the whole life cycle of SWM by offering the certain set of elements presented in Figure 6.2, which constitute a united SWM scheme.
- A model that envisages calculations of the "costs" and the "benefits" for each element scenario, which in itself.

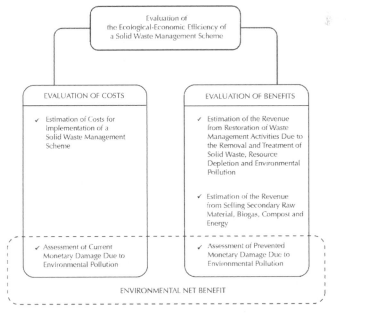

FIGURE 6.2 Procedures of the evaluation of the ecological-economic efficiency of a solid waste management scheme (by the authors) can be used as information to support decision making, and which can be integrated into uniform indices for the whole scheme.

According to Figure 6.2, the three key theoretical features of the proposed model are: (1) costs, (2) benefits, and (3) monetary damage effects:

(1) Regarding the cost part of the evaluation of a SWM scheme, it is necessary to take into account capital investments and all costs accompanying the implementation of a project. Taking into consideration the recommendations of the FCA methodology for MSW management (US EPA, 1997) and applying the "through" approach—which means going "through" or considering all costs involved in a scheme, rather than limiting the calculation to certain cost types—to the formation of the cost structure, this could be presented in general terms as:

$$C = \sum C_j = C_c + C_{op} + C_r + C_{en} + C_t + C_i + C_{ec} + C_{tax} + C_o \qquad (1)$$

where C = costs of implementation of a SWM scheme ending with landfilling; C_c = capital outlays; C_{op} = operating costs; C_r = costs for extensive and routine repairs; C_{en} = costs for creating engineering networks (infrastructure); C_t = costs for creating the transport scheme servicing a SWM scheme; C_i = costs for investment project services; C_{ec} = costs for current monetary damage caused by pollution of the environment; C_{tax} = cost for environmental taxes; and C_o =other costs. j = c, op, . . ., o.

In practice, the assessment of costs and benefits is based on, for example, information from the accounting system of the corporation in question and public data on the environmental situation of the region. Therefore, the finance departments of the companies or the civil servants of the authorities are supposed to be able to extract relevant and adequate data.

Additional information about the relative size of the studied scheme's costs could be useful for facilitating SWM decision making. Therefore, to provide an indicator based on generally applicable statistical facts for the size and extent of C in equation 1 compared with the total size of the average budget of a SWM actor of a certain kind according to available statistics, the authors suggest that equation 2 can express the size and extent of the current SWM scheme in relative terms:

$$R = C/TC_{avg} \qquad (2)$$

where R is the relative size and extent of the current SWM scheme of a certain kind, C $= \sum C_j$ is the same of the subcosts of the studied SWM actor of a certain kind, TC_{avg} is the total cost of the average SWM actor of a certain kind, and j = c, op, r, . . . , o.

Equation 2 is a mathematical indicator model representation based on the WAMED model expressing its implications for SWM scheme relativistic studies and is termed the company statistical business tool for environmental recovery (COSTBUSTER) indicator.

A new way of looking at waste is needed. Otherwise, the process of achieving an environmentally sound industry may be unacceptably slow. The paradigm shift that is argued for here involves equating industrial waste with normal products in terms of the allocation of revenues and costs, an approach that is termed the equality principle (Stenis, 2002). This approach forms the basis for the forthcoming discussion. The waste fractions studied are regarded as a company's output, which is mathematically considered in equation 3 below, and used for the allocation of revenues and costs to a certain waste fraction by multiplication of equation 3 with the costs or revenues in question that are to be allocated by splitting them up in their proper proportions:

$$PF = X/(Y+Z) \qquad (3)$$

where PF is the proportionality factor, X is the quantity of a certain waste fraction produced, Y is the quantity of normal product output, and Z is the sum of the quantities of all the different waste fractions produced. Of course, to apply equation 3, a suitable production or administrative unit must be defined, depending on the circumstances. Equation 3 represents the financial implications of the equality principle and is termed *the model for efficient use of resources for optimal production economy (EUROPE)*. (Stenis, 2005)

Here, it is proposed that C (the implementation cost of a SWM scheme) is allocated to the emissions from, for example, waste fires, from leachate, and from the odors of waste bales, multiplied by the proportionality factor (PF) in order to create economic incentives through inducing shadow prices for the reduction of such emissions (X = monetary damage value of a certain emission produced, Y = monetary value of the SWM scheme in normal operation, Z = monetary damage value of all the different emissions produced). The application in practice of the equality principle will be demonstrated in detail in another chapter (Pakhomova and Rikhter, 2001) that is closely related to this article. In particular, this latter article concerns SWM emission economics.

The FCA methodology provides a base for developing different concepts and tools for ecological and economic substantiation of a waste management activity, taking into account the range of this activity and the requirements of the national, regional, and local normative–legal base and the existing stereotypes. For most countries, the cost structure presented in Table 6.2 by Moutavtchi et al. (2008) is regarded as understandable and convenient for the practical implementation of SWM schemes. The suggested individual cost categories in the substantially extended cost structure of Table 6.2 in this study cover the majority of the occurring SWM cost items, the cost component being the other part of the basic CBA equation. Generally, the costs of major projects can be seriously understated. Considering this, it is important to perform sensitivity analysis (Pearce et al., 2006), and the reader is asked to bear in mind that, according to the authors, economics is not an exact science but merely a matter of providing useful tools to structure reality by the use of monetary entities. Thus, items in Table 6.2 such as "other costs" are justified to encapsulate miscellaneous expenses that sometimes cannot be foreseen. The allocation in Table 6.2 of costs for SWM facilities with shared use with whole plants (e.g., infrastructure), is preferably made based on the principles used within the accounting system of the current company.

(2) The transfer of benefits or values involves taking economic values from one context and applying them to another. Generally, analysts must rely on information from secondary sources (Kirkeby et al., 2006). Regarding the "benefits" part of the evaluation of a SWM scheme, in this study the payments received for waste removal, depletion of resources, waste treatment, and environmental pollution, as well as the benefits received from the sale of secondary raw materials, upgrading of materials to product,

compost, and other products of waste processing, are regarded as positively influencing the total economic profitability of a SWM scheme and, therefore, are used for estimation purposes. Thus, cost–revenue analysis in a business administration perspective is applied here together with CBA in an environmental impact perspective. This should be considered when estimations are made, for example, to obtain a decision basis for investments in additional environmental technology facilities.

During the comparison of various SWM schemes, one of the main criteria for the evaluation of economic efficiency is the degree of utilization. In this context, SWM should be considered as a component of the resource potential of a region (Stenis, 2002).

(3) In the WAMED model, evaluation of the ECO-EE of a SWM scheme carries with it an explicit assessment of the monetary damage that potentially arises through tangible effects such as the degradation of lands, pollution of surface and groundwater, pollution of the atmosphere, spreading of diseases (among the population and waste management personnel), and the disturbance of landscapes. The following procedures are proposed to enable calculation of the intangible monetary impact of a SWM scheme taking into account international requirements concerning the cost structure of an environmental protection project:

1. Calculation of *current* monetary impact caused by environmental pollution due to a SWM scheme. The effects can be determined through:

- Costs connected with the use of natural resources (remediation)
- Costs for elimination of the consequences of environmental pollution
- Indemnification of the population for loss of health

2. Calculation of *prevented* monetary impact caused by environmental pollution due to a SWM management scheme. The effects can be determined through:

- Costs for environmental pollution prevention, including environmental protection costs (construction and operating stages and post operating period) and costs for creation and functioning of a monitoring service
- Costs for prophylactic and protection measures to prevent the loss of population health and the health of personnel

TABLE 6.2   The cost structure for a solid waste management scheme in the waste managements' efficient decision (WAMED) model

	Cost category	Cost item
C_c	Capital outlays	• Costs for land acquisition (purchasing and preparation)
		• Costs for construction of main facilities, subsidiary industrial and service facilities, temporary buildings, and constructions
		• Costs for acquisition of buildings, premises, and constructions
		• Costs for acquisition of trucks and machinery and their setup
		• Costs for acquisition of intangible assets (e.g., know-how transfer, software, databases, patents, trademarks, licenses, know-how)
		• Tax and other obligatory payments on investment activities
		• Other capital outlays

TABLE 6.2 *(Continued)*

	Cost category	Cost item
C_{op}	Operating costs	• Incineration costs
		• Material expenditures
		• Energy costs
		• Costs for hazardous waste
		• Depreciation costs
		• Salaries of operating personnel
		• Tax and insurance costs
		• Rental payments
		• Administration costs
		• Costs for working environment of operating personnel
		• Costs for organization of work
		• Decommissioning costs (taking into account return of means from close-out sale) including retirement benefits of operating and service personnel
		• Other costs
C_r	Costs for extensive and routine repairs	• Costs for repair of buildings and constructions such as offices
		• Costs for repair of used and unused equipment, trucks, and machinery
		• Costs for repair of industrial premises
C_{en}	Costs for creating engineering networks (infrastructure)	• Costs for construction of water supply, sewage, power supply, gas supply, communication, telecommunication, and signaling facilities, e.g., cabling costs
		• Costs for hook-up of engineering networks such as connection fees
C_t	Costs for creating the transport scheme servicing a MSW management scheme	• Costs for construction of stationary transportation management facilities
		• Costs for construction of roads, including marking out and installation of means ensuring safety of traffic and pedestrians
C_i	Costs for investment project services	• Costs for research and design works
		• Costs for technical–economic substantiation, including costs for legal and public hearings
		• Costs for investigations related to a project and documentation process
		• Preliminary organization expenses (e.g., costs for registration, advertisement, capital issue, marketing, banking, and legal services)
		• Payment of long-term consulting and auditing services
		• Costs for scientific and engineering information and certification, for example, by ISO and EMAS
		• Costs for creation of a supply network
		• Costs for training and retraining of personnel
		• Other costs
C_{oc}	Costs for current monetary damage caused by pollution of the environment	• Costs related to remediation of use of natural resources
		• Costs for elimination of consequences of environmental pollution such as wind littering
		• Compensations of the population on loss of health including compensations of operating and service personnel
C_{tax}	Environmental taxes	Costs for environmental taxes due to current waste assortment grade and toxicity
C_o	Other costs	Other unanticipated costs including force majeure costs

3. In accordance with accounting methodologies accepted in most countries, the environmental protection costs are:

- Costs for acquisition, installation, maintenance, and repair of environmental protection equipment and the means of environmental control
- Costs for modernization with the purpose of ensuring the necessary level of environmental safety and resource saving
- Costs for implementation of environmental and resource-saving programs
- Costs related to management and control as regards environmental protection and use of natural resources

TABLE 6.3 The matrix of constituents of monetary damage caused by environmental pollution for a solid waste management scheme

Factors (j = 1 ... 5)	Element 1	Element i	Element n	Total damage
1. Estrangement of lands				$\sum_{i=}^{n} D_{i}$
2. Pollution of surface- and groundwater		$D_{ij} = D1_{ij} + D2_{ij} + D3_{ij}$		$\sum$
3. Polluti on of atmospheric air		$i = 1 \dots n$ $j = 1 \dots 5$		$\sum$
4. Spreading of disease (among the population)				$\sum$
5. Disturbanc e of lan scapes				$\sum$
Total damage	$\sum$	$\sum_{j=}^{n} D_{ij}$	$\sum_{j=}^{n} D_{nj}$	$\sum_{i=1}^{n}\sum_{i=1}^{n} D_{ij}$

$D1_{ij}$, material damage; $D2_{ij}$, damage to the health and living ac tivities of the population; $D3_{ij}$, damage to resources and industries

The calculation of the monetary values of, in particular, the current and also the prevented economic damages due to environmental emissions may be based on estimations of the costs for elimination, remediation, and prevention of pollution. More precisely, the specific amounts may be obtained through applying historical and budgeted accounting data for oil-polluted soils and collection and treatment of landfill gas and leachate, for example, according to the method of evaluation presented by Moutavtchi et al. (2008). Alternatively, the approach introduced by Stenis et al. (2005) can be used to obtain estimates of the monetary values of emissions (sol) from accidental open burning, pollution (liq) from leachate and emissions (g) from odors at a SWM scheme. When applying the latter approach, here also the accounting system of the company in question is utilized to obtain precise amounts to enter into the WAMED model that in this case is applied to provide *shadow costs* to additionally allocate to different kinds of pollution. Each SWM scheme has a certain negative potential impact on the environment and the health of the population.

Thus, during the substantiation of a specific variant of a SWM scheme or development of a complex technological solution, a required stage is the comparative analysis of current and prevented monetary damage effects caused by environmental pollution, that is, the evaluation of the ECO-EE of a scheme. The matrix in Table 6.3 constitutes the guiding principle of a general method based on mathematics for obtaining the necessary information for assessment of the monetary damage by summation of the separate impacts of different elements or schemes. The single values to be entered into Table 6.3 are estimated by the utilization of historical and budgeted data from the accounting system of the company in question according to the approach proposed by Moutavtchi et al. (2008) as regards current and prevented damage (Cec) or by allocating *shadow costs* according to the findings by Stenis et al. (Solid waste management Baling scheme economics; a Swedish case study, 2009, submitted).

This matrix could be used in decision support systems for:

- Calculation of the total ECO-EE of the entire SWM scheme
- Evaluation of the environmental impact of the chosen elements of a SWM scheme and selection of "hot spots" and priorities for the scheme
- Substantiation of the selection of the implementation variants of the SWM scheme stages considering the ECO-EE
- Optimization of a SWM scheme according to the greatest ECO-EE

6.5 DETERMINATION OF BENEFITS OF THE ECO-EE DURING EVALUATION

The most difficult part of the practical application of CBA is to determine the tangible and intangible benefits or effects of the ECO-EE of a SWM scheme. The difficulty of determining the effects is caused, *inter alia*, by the following factors:

- The problem of predicting the dynamics of prices, for instance, on the recyclables market and monetary indices in the future, which complicates the cost calculation procedure in CBA
- The problem of including the intangible effects, such as social results, an increase in the aesthetic value of a landscape, and the employment growth for the residential population when implementing a SWM scheme, into the calculations, given the difficulty of representing these intangibles directly in monetary terms
- The problem of assessing the monetary damage caused by pollution of the environment

The authors propose that during evaluation of the ECO-EE of a SWM scheme, the following benefits should be taken into account:

- Revenues due to waste removal, disposal, and treatment
- Revenues from selling secondary raw material, compost, and other products of waste processing

Prevention of damage to the environment and the health of the population that can be assessed by means of:

- Costs for preventing pollution of the environment, including environmental protection measures during the construction and operating stages and the post operating period and costs for the creation and functioning of a monitoring service
- Costs for prophylactic measures to prevent losses of the health of the population and waste management personnel, which are likely to lead to an increase in the working efficiency of the population and personnel and an extension of the employable period
- Costs for public information material and meetings to prevent public fear of pollution generation

During evaluation of the ECO-EE of a SWM scheme, it is necessary to take into consideration the fact that payments of remunerations received for waste removal, waste treatment, and avoidance of pollution of the environment, as well as revenue received from the sale of secondary raw materials, compost, and other products of waste processing, can positively influence the total economic profitability of a SWM scheme. However, it is possible only when real prices are paid for "products" such as MSW, secondary resources, and recyclables. Presently, payments for pollution do not adequately cover common needs for investments in environmental protection activities. The main point is that when there is success in the development of a reasonable charge—with respect to any user of natural resources—for the pollution of the environment, then the calculation turns out to be too complicated for practical application. When there is a simple and easily applied calculation, the payment does not take into account all aspects of pollution. Therefore, the collation of existing and necessary payments should be carried out during the evaluation of the ECO-EE of SWM schemes (Pakhomova and Rikhter, 2001). The process of the waste managements' decision making, described above, was previously summarized by Moutavtchi et al. (2008).

6.6   SWM SCHEME CASE STUDY

Before considering a case study involving landfilling as a treatment option, it should be mentioned that according to the regulations, incineration is prioritized in SWM, with only inorganic tails of waste to be landfilled (Council of the European Union, 1999). The theoretical basis of the "cost" approach to evaluate the ECO-EE of SWM schemes, according to the proposed WAMED model concept, enables evaluation of the cost items in the profit and loss account of a fictive SWM scheme based on real world Swedish data for the year 2002.

The results of the evaluation are presented in Table 6.4. The basic principles for the practical usefulness of the WAMED model and the COSTBUSTER indicator model are evaluated based on the analysis of the realism in the estimation.

The case that is analyzed is typical for a plant handling approx. 40,000 tons MSW per year. The data used represent the common and essential components of a typical scheme, and this fact leads to certain terms in the basic model (equation 1) necessarily being omitted. The applied 1-year time horizon is suitable to obtain an instant comprehension of the current cost status of a scheme that is useful for extrapolation over longer time spans. The internal rate of return (IRR) is regarded as irrelevant, the current project being the only option. Generally, the IRR is agreed not to be used for ranking and selecting mutually exclusive projects. The discount rate is assumed to be constant, this being consistent with intergenerational fairness (Pearce et al., 2006).

Application of the WAMED model, according to equation 1, to the fictive SWM scheme for the landfill, gives the following estimation (k€):

$$C = \sum_j C_j = C_c + C_{op} + C_r + C_i + C_{ec} \tag{4}$$

where Table 6.4 gives:

C = costs for implementation of the SWM scheme in 2002 = 4775,

C_c = capital outlays = 300 + 150 + 50 + 100 + 25 = 625,

C_{op} = operating costs = 575 + 50 + 50 + 25 + 175 + 325 + 575 + 875 + 150 + 450 + 100 + 25 + 25 + 125 + 25 + 175 + 75 = 3800,

C_r = costs for repairs = 125,
C_i = costs for investment project services = 50 + 25 + 25 + 25 = 125,
C_{ec} = costs for current damage = 100.
Application of the COSTBUSTER indicator to the fictive SWM scheme 2002 yields:

$$R = \sum C_j / TC_{avg} \qquad (5)$$

j = c, op, r, i, ec,

where $\sum C_j$ is the current annual total cost of the fictive landfill in 2002 = k€ 4775, TCavg is the current annual total cost of the average Swedish municipal landfill actor = €71/ton at a fill rate of 40,000 tons/year = k€2840 (Hogg, 2006). Equation 5 gives, R, the relative size and extent of the current SWM scheme, as k€4775/k€2840 = 1.68 = 168% of the average Swedish municipal landfill actor.

6.7 DISCUSSION AND RESULTS

In the process of designing the WAMED model, a number of models with CBA elements were useful sources of ideas, even though an estimation of the monetary damage to the environment is not yet offered by these models in an explicit monetary form. Its novelty is a substantial scientific contribution of the WAMED model, as well as the possibility of its fruitful combination with the *equality principle* in order to reduce the existence of waste management-related emissions.

In particular, the Waste Plan model contributed through the introduction of the FCA approach to the model through its emphasis on accounting for and allocation of all the costs for SWM. Regarding the development of a SWM scheme for the WAMED model, the main contribution was in the form of the approach used in the IWM-2 model, featuring operation costs as costs incurred when implementing a scheme. The rest of the models studied (Table 6.1) mainly provided general tools for the current model build up, for example, WISARD5 provided a useful framework of cost items, as did EPIC/CSR to a certain extent. Other models listed in Table 6.2 are regarded as less useful for the purpose of contributing to the design of the WAMED model. Thus, the WAMED model can be said to represent an extension and, to a certain extent, a synthesis of similar existing models.

TABLE 6.4 Analysis of costs (thousand 3) for a fi ctive Swedish landfill (approximate data for 2002)

Cost component	Cost item	MSW collection	MSW transporta-tion	MSW in-cineration, landfilling
Capital outlays	Implements acquisition	300		150
C_c	Interest costs	50		100
	Other	25		
	Total	625		
Operating costs (annual)	Operation of vehicles		575	
C_{op}	Maintenance of containers	50		
	Maintenance of containers at collection stations	50		
	Operating costs for collection stations	25		

TABLE 6.4 *(Continued)*

Cost component	Cost item	MSW collection	MSW transportation	MSW incineration, landfilling
	Operating costs for collection, including costs for	175		
	source separation (newspapers, colored glass)	325		
	Costs for MSW landfilling			575
	Salaries to the personnel	875		
	Salaries to managers	150		
	Administration costs	450		100
	Other costs	25		
	Total	3375		
Costs for repair	Costs for repair and maintenance of the garage		125	
C_r	Total		125	
Costs for investment project services	Information services	50		25
	Associated services			25
C_i	Other costs	25		
	Total	125		
Current damage (remediation costs)	Remediation of oil-polluted soils 100			
	Total		100	
C_{oc}				
Prevented damage (costs for preventing pollution)	Collection and treatment of landfill gas			25
	Collection and treatment of leachate			125
	Operations at the old landfill			25
C_{op}, C_i	Treatment of oil-contaminated waste			175
	Costs for research	75		
	Total	425		

MSW, municipal solid waste

By applying the outcome of the literature study in the Introduction, it can be concluded that the current scientific gap filled by the WAMED model is mainly the need to provide practically useful tools for SWM based on a methodology that enables expression of the current decision basis in monetary terms. In particular, the suggested employment of common economic tools related to business administration is featured in the introduced models.

Depending on the time horizon applied, the aspect of sustainability is considered. Usually, the long-term perspective is more connected to the concept of sustainability. That is also the reason why the chosen definition of the ECO-EE is plausible. The practical applicability and the promotion of sustainability integrated into the WAMED model is mirrored by the time perspective applied for decision making at the current corporate, municipal, and/or regional levels.

Historically, the WAMED model fits well with the timeline of model design because it is consistent with recent developments that regard damage assessment as an important part of contemporary environmental management. Its main contribution to the knowledge accumulation process can be found by carrying out comparative analysis of current and prevented monetary damage effects caused by environmental pollution, that is analysis of the ecological efficiency expressed in *monetary* terms as regards the implementation of a SWM scheme. Therefore, the WAMED model can be expected to provide an output that is more directed towards environmental matters, better adapted to business administration-related demands, and presented in a more easy-to-grasp way than similar prevailing models. Thus, the major advantage of the WAMED model is that, for academics, civil servants, and businessmen, it is likely to facilitate practical SWM decision making better than similar existing models do.

The WAMED model has been shown to produce useful and reasonable results when applied to the current case study SWM scheme in order to evaluate its ECO-EE. This is confirmed by the application of the COSTBUSTER model, based on the WAMED model outcome, to determine the relative size and extent of the management scheme in question, pointing in the direction of a reasonable current extra 68%. This figure enables the investor in question to know that his/her intended investment, in relative terms, is comprehensible when performing, for example, project investment appraisals. If its relative size had been, let us say, three times as big as the average plant, the potential investor might have become somewhat suspicious and question whether he/she really should embark on such a bold endeavor.

This study shows that, in general, the WAMED model's practical usefulness is good owing to its rather simple mathematical approach, which is based on a statistically acceptable foundation, and to the fact that it derives its sources from the municipal waste management sector. Therefore, possible applications are to be found in municipal and regional planning. The prospects for the future of the model are regarded as good because it satisfies several features applicable to different SWM actors.

The principal requirements for the success and the practical applicability of evaluation methodologies for the ECO-EE of SWM schemes should be: (i) selection and substantiation of those benefits for implementation of a SWM scheme, which should be included in the analysis; (ii) measuring results quantitatively and in monetary terms; (iii) the availability of necessary market information; and (iv) competent data treatment, with the help of statistical and econometric models.

The selection process for a variant of a SWM scheme implies carrying out the following:

1. The aggregation of the proposed calculation procedures
2. The generation of new alternative solutions to the problem in question
3. Formulation of the selection criteria considering certain views on values and priorities
4. The selection of an optimal variant solution

In Sweden nowadays, small and medium-sized waste management companies, in particular, struggle with a problematic profitability situation. Often, they face the necessity of deciding whether to operate on their own or to join forces with mainly municipal waste management actors. Therefore, the models presented here are most useful as management tools to provide the basis for these kinds of decisions. By using the WAMED model concept to define the total corporate cost structure, managers can improve control over their financial situation compared with what would otherwise be the case when, for example, applying cost-revenue analysis for project investment appraisal purposes.

Naturally, however, not all efforts to evaluate the environmental impact of a SWM scheme can be expressed in monetary terms. For example, intangible effects such as the impact of political environmental ambitions and pressure from lobby groups are difficult to quantify.

Likewise, it is difficult to quantify people's attitudes and their understanding of and willingness to accept and correctly apply environmentally related models, including the WAMED model.

Nevertheless, the WAMED model constitutes a means of recognizing environmental realities in a way that is easy to cope with. The major selling feature of the model is its simplicity compared with most models with CBA elements. In addition, when applied in practice, the WAMED model is likely to show greater applicability regarding cost accounting.

This is an advantage over existing models, including, for example, IWM-2, WastePlan, and EASEWASTE, which are more complicated in this respect or do not have applicability to cost accounting at all. The application of the previously developed EUROPE model in connection with the WAMED model enables management to pinpoint, through internal economic incentives, unwanted environmental phenomena related to the operation of a SWM scheme. Potential model users include practitioners, academics, and others desirous of estimating the monetary impacts and the tangible effects of SWM schemes and those wanting to reduce their related environmental pollution.

Summarized, the WAMED model developed for evaluation of the ECO-EE of a SWM scheme encompasses the following features:

- It considers the entire SWM scheme (Figure 6.2).
- It reflects an integrated approach to solving the problem of simultaneously decreasing the negative impacts of SWM on the environment and the health of the population while increasing the economic benefits of SWM projects (including those in the field of environmental protection and rational use of natural resources).
- It can be considered as a unified and adaptable information support tool for decision making by SWM actors at the corporate, municipal, and regional levels. It is important to recognize that the methodological provisions of WAMED are based on CBA.
- It provides an information support tool for SWM decision making at the corporate, municipal, and regional levels to improve small- and medium-sized company competitiveness in particular.
- It uses the "through approach" for estimating the implementation costs of a SWM scheme to optimize control over economic benefits that are achieved.
- It enables comparative analysis of current and prevented monetary damage effects to be carried out when implementing a scheme.
- It is possible to combine it with the EUROPE model based on the *equality principle* to create economic incentives to reduce waste management-related emissions.

The validity of the present study was ascertained through the application of recognized models as a basis for the design of the WAMED model and, as a consequence, the COSTBUSTER indicator model. The combination of the WAMED model with the extensive use of empirical data as inputs ensures the model's validity. The reliability of the models applied here was ascertained by the realistic outcome of the case study based on relevant facts taken from the daily reality of SWM. The case study hence showed reliable and trustworthy results.

Further research will focus on algorithm development based on existing techniques for assessment of monetary damage caused by environmental pollution in the form of tangible effects such as the estrangement of land, pollution of surface and ground-water, pollution of the atmosphere, spreading of disease (among the population), and the disturbance of landscapes. In addition, a set of indicators for determination of the "consequences" of waste management technologies will be developed. Also, studies will be made of how to use the methodology for measurement of environmental footprints and calculation of carbon dioxide taxes. Finally, further research with macroeconomic extensions will be conducted to determine how to naturally encompass regional and global cost aspects into the models in order to constantly improve them based on an ever-increasing knowledge of our living environment.

6.8   CONCLUSIONS

The following conclusions were reached in the course of applying the WAMED model and the COSTBUSTER indicator model to the landfilling concept currently applied in Sweden and hence studied in the case study:

- It was possible to estimate the cost part of the total monetary value of the scheme in the short term.
- The research presented in this article, aimed at information support for SWM decision making, was useful when focusing on the practical aspects of CBA for SWM.
- The case study investigated the practical application of the theoretical provisions of CBA to a Swedish landfill and was useful for evaluation of the ECO-EE of the proposed SWM scheme.
- It is proposed:
- To use the "through" approach for estimating the implementation costs of a SWM scheme.
- To assess the monetary damage caused by environmental pollution which appears in the form of tangible effects such as the estrangement of lands, pollution of surface and groundwater, pollution of the atmosphere, the spreading of disease (among the population and waste management personnel), and the disturbance of landscapes.
- To carry out comparative analysis of the current and prevented monetary damage when implementing a SWM scheme.
- To carry out comparative analysis of the cost, benefit, and economic revenue components of a SWM scheme.
- To combine the WAMED model and the COSTBUSTER indicator model with the EUROPE model based on the *equality principle.*

It is concluded that the future relevance of the WAMED model for practical waste management is good because its rather simple mathematical approach is applicable to different SWM actors and is based on a statistically acceptable foundation. Generally, the users of the model are expected to be potential investors in SWM plants and those who wish to estimate the impact of applying a SWM scheme in monetary terms and to reduce its related environmental pollution.

Based on the analysis performed, the recommendations are to:

- apply the WAMED model to SWM schemes in order toevaluate their ECO-EE
- apply the COSTBUSTER model, based on the current WAMED model outcome, to determine the relative size and extent of SWM schemes
- study how to combine the WAMED model and the COSTBUSTER indicator model with the EUROPE model based on the *equality principle*

ACKNOWLEDGMENT

The authors gratefully acknowledge the support of the Swedish Institute (Svenska Institutet) and the Kalmar Research and Development Foundation—Graninge Foundation (Kalmar kommuns forsknings—och utvecklingsstiftelse—Graninge-stiftelsen).

The authors would also like to thank Dr. Diauddin Nammari for providing input in the preparation of the article.

KEYWORDS

- **Cost-benefit analysis**
- **Decision making**
- **Ecological–economic efficiency**
- **Full cost accounting**
- **Solid waste management**

REFERENCES

Barlishen, K. D. and Baetz, B. W. Development of a decision support system for municipal solid waste management systems planning. *Waste Manag. Res.*, **14**, 71–86 (1996).

Berger, C., Chauny, F., Langevin, A., Loulou, R., Riopel, C., Savard, G., and Waaub, J. P. EUGENE: An optimization-based decision support system for long-term integrated regional solid waste management planning. In Proceedings of the International Workshop on Systems Engineering Models for Waste Management. Swedish Environmental Protection Agency, Stockholm, pp. 11–29 (1998).

Björklund, A. *Environmental system analysis of waste managementexperiences from applications of the ORWARE model*. Doctoral thesis, TRITA-KET-IM 2000:15. Division of Industrial Ecology, Department of Chemical Engineering and Technology, Royal Institute of Technology, Stockholm (2000).

Chang, N. B., Shoemaker, C. A., and Schuler, R. E. Solid waste management system analysis with air pollution and leachate impact limitations. *Waste Manage. Res.*, **14**, 463–481 (1996).

Chang, N. B., Yang, Y. C., and Wang, S. F. Solid waste management system analysis with noise control and traffic congestion limitations. *J. Environ. Eng.*. **122**(2), 122–131 (1996).

Council of the European Communities. Council Directive 91/689/EEC of 12 December, 1991 on hazardous waste. *Off. J.*, **L377**, 31/12/1991, 0020–0027 (1991).

Council of the European Union. Council Directive 1999/31/EC of 26 April, 1999 on the landfill of waste. *Off. J. Euro. Comm.*, **L182**, 16/07/199, 0001–0019 (1999).

EPIC and CSR. *Integrated solid waste management tools: User guidance document*. Environmental and Plastics Industry Council, Corporations Supporting Recycling, Ontario (2000).

Eriksson, O. *A systems perspective of waste and energy: Strengths and weaknesses of the ORWARE model*. Licentiate thesis, TRITA-KET-IM 2000:16. Division of Industrial Ecology, Department of Chemical Engineering and Technology, Royal Institute of Technology, Stockholm (2000).

European Commission. *Cost-benefit analysis of the different municipal solid waste management systems: Objectives and instruments for the year 2000*. Office for Official Publications of the European Communities, European Commission, Luxembourg (1996).

European Parliament and the Council of the European Union. Directive 2000/76/EC of the European Parliament and of the Council of 4 December, 2000 on the incineration of waste. *Off. J. Euro. Comm.*, **L332**, 28/12/2000, 0091–0111 (2000).

Everett, J. W. and Modak, A. R. Optimal regional scheduling of solid waste systems. I: Model development. *J. Environ. Eng.*, **122**(9), 785–792 (1996).

Finnveden, G., Björklund, A., Ekvall, T., and Moberg, Å. *Models for waste management: Possibilities and limitations*. Environmental strategies research. Royal Institute of Technology, Stockholm (2006).

Finnveden, G., Johansson, J., Lind, P., and Moberg, Å. *Life-cycle assessments of energy from solid waste*. FOA Report No FOA-B-00-00622-222-SE, FMS Report No 137, Swedish Defence Research Agency, Stockholm (2000).

Gielen, D. J. *The MARKAL system engineering model for waste management*. In Proceedings of the International Workshop on Systems Engineering Models for Waste Management. Swedish Environmental Protection Agency, Stockholm, pp. 31–51 (1998).

Goldstein, J. and, Sieber, J. *Waste Plan: Software for integrated solid waste planning*. User guide for version 5.0. Tellus Institute, Boston, MA (2003).

Gottinger, H. W. *Economic model and applications of solid waste management*. Gordon and Breach, London (1991).

Groom, M. J., Meffe, G. K., and Carroll, C. R. *Principles of conservation biology*, 3rd ed. Sinauer, Sunderland, CT (2005).

Hanley, N. Cost-benefit analysis of environmental policy and management. In Handbook of *Environmental and resource economics*. Jeroen, C. J. M. and van den Bergh (Eds.) Edward Elgar, Cheltenham, pp. 824–836 (1999).

Hellweg, S., Binder, M., and Hungerbühler, K. Model for an environmental evaluation of waste treatment processes with the help of life-cycle assessment. In *Proceedings of the International Workshop on Systems Engineering Models for Waste Management*. Swedish Environmental Protection Agency, Stockholm, pp. 241–251 (1998).

Hogg, D. *Costs for municipal waste management in the EU*. Eunomia Research and Consulting, Available at: http://ec.europa.eu/environment/waste/studies/pdf/eucostwaste.pdf (accessed March 27, 2010) (2006).

Kijak, R. and Moy, D. A decision support framework for sustainable waste management. *J. Ind. Ecol.*, **8**(3), 33–50 (2004).

Kirkeby, J. T., Birgisdottir, H., Hansen, T. L., Christensen, T. H., Bhander, G. S., and Hauschild, M. Environmental assessment of solid waste systems and technologies: EASEWASTE. *Waste Manag. Res.*, **24**, 3–15 (2006).

Ljunggren, M. *The MWS modela systems engineering approach to national solid waste management*. In Proceedings of the International Workshop on Systems Engineering Models for Waste Management. Swedish Environmental Protection Agency, Stockholm, pp. 97–119 (1998).

McDougall, F. R., White, P. R., Franke, M., and Hindle, P. *Integrated solid waste management: A life cycle inventory*. Blackwell Science, Oxford (2001).

Morrissey, A. J. and Browne, J. Waste management models and their application to sustainable waste management. *Waste Manag.,* **24**, 297–308 (2004).

Moutavtchi, M., Stenis, J., Hogland, W., Shepeleva, A., and Andersson, H. Application of the WAMED model to landfilling. *J. Mater. Cycles. Waste. Manag.,* **10**, 62–70 (2008).

Organization for Economic Cooperation and Development. Core set of indicators for environmental performance reviews: A synthesis report by the group on the state of the environment. *Environment monograph* no. 83. OECD, Paris (1993).

Pakhomova, N. V. and Rikhter, K. K. *Economics of natural resources use and environment protection* (in Russian). Publishing House of St. Petersburg State University, St. Petersburg (2001).

Pearce, D., Atkinson, G., and Mourato, S. *Cost-benefit analysis and the environment: Recent developments.* OECD, Paris (2006).

PWC and URS. *Life-cycle tool for waste management in New Zealand. WISARD* reference guide. PricewaterhouseCoopers (PWC), URS New Zealand, Auckland (2001).

Roberge, H. D. and Baetz, B. W. Optimization modeling for industrial waste reduction planning. *Waste Manag.* **14**(1), 35–48 (1994).

Solano, E., Dumas, R. D., Harrison, K. W., Ranjithan, S. R., Barlaz, M. A., and Brill, E. D. Life-cycle-based solid waste management. II: Illustrative applications. *J. Environ. Eng.,* **128**(10), 993–1005 (2002).

Solano, E., Ranjithan, S. R., Barlaz, M. A., and Brill, E. D. Life-cyclebased solid waste management. I: Model development. *J. Environ. Eng.,* **128**(10), 981–992 (2002).

Stenis, J. Industrial waste management modelsa theoretical approach. Licentiate dissertation. Department of Construction and Architecture, Division of Construction Management, School of Engineering, Lund University, Lund (2002).

Stenis, J. Industrial management models with emphasis on construction waste. Doctoral thesis. Department of Construction and Architecture, Lund Institute of Technology, Lund University, Lund (2005).

Sundberg, J. *MIMES/wastea systems engineering model for the strategic planning of regional waste management systems.* In Proceedings of the International Workshop on Systems Engineering Models for Waste Management. Swedish Environmental Protection Agency, Stockholm, pp. 85–95 (1998).

Tsilemou, K. and Panagiotakopoulos, D. Approximate cost functions for solid waste treatment facilities. *Waste Manag. Res.,* **24**, 310–322 (2006).

US EPA. *Industrial waste management evaluation model (IWEM). User's guide.* EPA Report No 530-R-02–013. Office of Solid Waste and Emergency Response (5305W), United States Environmental Protection Agency, Washington, DC (2000).

US EPA. Full cost accounting for municipal solid waste management: A handbook. EPA Report No 530-R-9 5-041. United States Environmental Protection Agency, Washington, DC (1997).

Vigsø, D. Deposits on single-use containersa social cost–benefit analysis of the Danish deposit system for single-use drink containers. *Waste Manag. Res.,* **22**, 477–487 (2004).

7 A Model for Solid Waste Management in Anambra State, Nigeria

Otti V. I.

CONTENTS

7.1   INTRODUCTION

Indiscriminate dumping of solid wastes along the streets and roads corners causes a lot of deadly infectious diseases which could be responsible for the large proportion of morbidity and mortality in Nigeria. A deterministic model needed for short and long term waste management and management information system in Anambra State sanitation and environmental protection agency (ANSEPA) is considered in this chapter. A review of literature on model methods is presented, with brief method of the study and analysis used for the determination of the required results. Moreover, this study was aimed to determine which type of integrated solid waste management option or program will be used to implement minimized cost and maximized benefit (benefit cost ratio) over a long period of planning period. Consequently, the model will be used by the decision ma kers in finding the solution to environmental, economical, sanitary, technical, and social goals, through the use of equipment, routine maintenance, personal and sundry.

Solid waste is a system of engineering, involving substantial engineering content, that is particularly set for actions which will best accomplish the overall objectives of the decision makers, within the constraints of law, morality, economics, resources, political, and social pressure and which will govern the physical life and other natural sciences solid waste management is defined as the discipline associated with the control of generation, storage, collection, transfer and transport, processing, and disposal of solid waste in a manner that is in accord with the best principles of public health, economic, engineering, conservation, aesthetics, and other environment consideration that is also responsive to public attitudes (Tchobanoglous et. al., 1993).

Waste management in the three urban cities of Anambra State, namely Awka, Onitsha and Nnewi and some few local governments is becoming an increasing problem daily and a complex task. The ANSEPA are being considered as the base scenario for development of this waste management model. The State Government has a major waste management issue and has been noticeable since the 1980s. The waste management which in the past times has been addressed with various methods by different administrations in tackling the waste problem yielded nothing.

However, the Board of ANSEPA is charged with the following responsibilities:

1. Removal, collection and disposal of domestic commercial and industrial generated waste.
2. Cleaning and maintenance of Public drainage facilities
3. Cleaning streets of Awka, Onitsha, and Nnewi.
4. Removal and disposal of abandoned Scrapped vehicles
5. Streets sweeping of major roads.

In the operational period of the waste disposal board, its activities were limited to the three urban towns and just recently, due to population growth and progressive urbanization, the service areas are expanded to some few local government areas.

At all times, human activities have generated waste in various forms in gaseous (abattoirs), liquid and solid. These wastes have often been discarded because they were all considered as negative value goods. The more prevalent method of disposal of these wastes have been to first collect them from their source and then burn them in a landfill site or throw them in the surrounding deep erosion gullies in the state.

However, the steady increase of landfill site, deposition in the gullies, and waste generally has caused a lot of havoc to the potable water being extracted from downstream and ground water. Currently, the emergence and development of new public environmental consciousness have created a strong negative attitude toward landfill and deposition into gullies.

The national and state regulation to protect the environment have increased the cost of developing new landfill and deposition in the gullies, also siting has become increasingly difficult because the public oppose having such facilities nearby. Solid waste management has become a major concern in industrialized developing countries, like Nigeria. The ideal way to improve the situation would be, to reduce the generation of waste. But contrarily, this goes against the people's will to preserve their life style and thus to consume more food.

Consequently, the society is searching for improved method of waste management and ways to reduce the amount of waste material which can be reused or transformed into useful material (e.g., plastic and some cast iron from sites–mechanics) if managed properly. Many waste management options have been proposed previously by some committees set up by Government but has always been poorly implemented which resulted in failure, unproductiveness and corruption. The most, the implementing agency, cannot foresee or properly forecast the out-come of such program, and also properly and well planned scheduling processes were not included in the management system and corruption which has eaten deeply in the system.

Different waste management options must be combined intelligently in a way as to reduce the environmental and social impact at an acceptable cost for the masses in the state. This combined option is called integrated solid waste management and system approach should be used for the assessment of the competing option.

7.2 OBJECTIVE

An integrated waste rearranged system of plan must place an emphasis not only on which specific waste management option are to be chosen, but on the scheduling of these location of facilities and equipment (Tippers, Pail loader, and Bulldozer).

A more flexible choice and scheduling program for waste management options which must be, also to adapt to changing conditions need to be considered in the plan. The basic aim here is to allow decision makers to be able to determine the optimal times to implement and discontinue or close the waste management program and facilities. Throughout the planning period, this should include a determinist schedule plan of when and what recycling program to implement and the landfill is to be opened or gully to be filled in a given planning period. Also the schedule option should minimize the overall cost associated with the solid system for a defined planning period. This is achieved by integrating a cost minimization, example minimizing the cost of equipment maintenance (Bulldozers, Pail loaders, and Tippers).

Some operation research model is particularly well suited for the description of complex task method involving some variables as constraints (Equipment: Bulldozer, Pailoaders, and Tippers). These models may be used to help understand the complexity of the system as well as assessing the long term role and impact of the new technology option, Gottinger, (1986) in his integrated model of waste optimization proposed a network model which would help decision makers in the waste management and facility sitting decision. Also, Kaila (1987) developed a model for the strategic evaluation of municipal solid waste management system.

7.3 METHODOLOGY

Optimization model for solid waste management system engineering approach to planning, scheduling, cost minimizing, maintenance and general management of solid waste management system, serves as a control tool for decision management makers in the areas of waste management (Mackenzie and David, 1998). The necessity for this system approach lies in the fact that waste management in recent times have developed to a complex task. The system of optimal model is focused on the ANSEPA, as a means of eradicating waste littering along the streets and roads and that concerns municipal and local waste management system. The optimal system represents a group of specific municipal and local system and is defined in a set of existing an optimal treatment process and flow. The input data for the system is a sum of the specific system of the category; both the municipal and local waste management systems are represented in the model. The compliance between the system model representing state government waste transport and the mathematical representation of the model is set up from a number of standardized devices defined in the model (Sundberg, 1993); each device corresponds as described accordingly by the relations between input and output flows of material.

The material waste is modeled by a number of factions- Plastic, glasses, personnel, purchase, and maintenance of equipment. The example, is that if ANSEPA Embarks on a massive environment project called "Operation Sweep All and Clean Up" in the state using an optimization model system (Linear program) by effective use of Pail loader, Tipper, and Bulldozer in its work of environmental sanitation, the cost of purchasing one unit of Pail loader, Tipper and Bulldozer are N20 million, N10 million, and N30 million, respectively

TABLE 7.1 Linear programming-maximizing result

	x_1	x_2	x_3	S_1	S_2	S_3	Z
	-7	-5	-8	0	0	0	0
S_1	2	1	3	1	0	0	100
S_2	4	3	3	0	1	0	240
S_3	1	1	2	0	1	1	86
	$-1/3$	0	-3	0	5/3	0	400
S_1	2/3	0	2	1	$-1/3$	0	20
x_2	4/3	1	1	0	1/3	0	80
S_3	$-1/3$	0	1	0	$-1/3$	1	6
	0	0	-2	½	2/3	0	410
x_1	1	0	3	3/2	$-1/2$	0	320
x_2	0	1	-3	-2	1	0	40
s_3	0	0	-2	½	$-1/2$	1	16
	0	0	0	1	1	1	426
x_1	1	0	0	3/4	$-1/2$	$-3/2$	6
x_1	0	1	0	$-5/4$	1/4	3/2	64
x_3	0	0	1	½	$-1/4$	½	8

Therefore $x_1 = 6$, $x_2 = 64$, $x_3 = 8$, $Z = 426$. Number of equipment and amount spent; Pail loader = 6 × 7 = N 42; Tipper = 64 × 5 = N 320; Bulldozer = 8 × 8 = N 64; Total cost = N426 million = USD 2.84 million.

The routine maintenance cost for each Pail loader, Tipper, Bulldozer are N4,00,000, N3,00,000, and N3,00,000, respectively. Personnel and Sundry cost for running each are respectively N1,00,000, N1,00,000, and N2,00,000. The maximum allowable budget of the Authority (Agency) for the personnel sundry, purchase, and maintenance has been determined, that in relative terms the benefit (in term of clearing waste) derived from the use of each of the equipment above is in the ratio of 7:5:8. Therefore to maximize, the environmental benefit, which is the main aim of the agency, the number of equipment; Pail loader, Tipper and Bulldozer should be determined using linear programming.

The model takes into account in the scheduling decision, benefit overtime, budget constraints, and constraints on the number of equipment available to effectively implement the project. Moreover; decision making is a vital tool for the engineer, in relation to planning, design, execution or maintenance (Peavy et. al., 1985).

7.4 DISCUSSION AND RESULTS

7.4.1 Optimal Solution to Sanitation Problem

7.4.1.1 *Maximizing the Results of Good Environmental Cleanliness*
In 1979, Tomas et. al. stated that Linear Programming is an objective function that optimizes cost or gain as it is subjected to the constraints and involves some decisions (Table 7.1).

Maxi. $Z = 7x_1 + 5x_2 + 8x_3$ (Benefit point)
Subject: $2x_1 + x_2 + 3x_3 \leq 100$ (Personal and sundry)
$4x_1 + 3x_2 + 3x_3 \leq 240$ (Equipment Purchase)
$x_1 + x_2 + 2x_3 \leq 86$ (Routine Maintenance) $x_1 + x_2 + x_3 \geq 0$ and integer

Standard form
$Z = 7x_1 + 5x_2 + 8x_3$
Subject to: $2x_1 + x_2 + 3x_3 \leq 100$
$4x_1 + 3x_2 + 3x_3 \leq 240$
$x_1 + x_2 + 2x_3 \leq 86$

Inclusion of non-basic variable and basic variable
$Z = 7x_1 + 5x_2 + 8x_3 + 0S_1 + 0S_2 + 0S_3$
Subject to: $2x_1 + x_2 + 3x_3 + 0S_1 + 0S_2 + 0S_3$
$4x_1 + 3x_2 + 3x_3 + 0S_1 + 0S_2 + 0S_3$
$x_1 + x_2 + 2x_3 + 0S_1 + 0S_2 + 0S_3$

The development of optimization model and execution process are ordered and streamlined to effectively achieve the required result, as in the determination of required result to be addressed by the model and area of focus in implementation (Kaila, 1987). This was done first to determine the scope of the design and to ensure a necessary guideline for the project work with the full aim of achieving a competitive result even both in analysis design and work.

Also, it determined planning models for project execution, which consist of planning of models and modules needed for execution of the model. Among the major purpose of this model, is the role it plays in economics development, via high level of economics productivity and stimulate immediate and rapid growth regards to employment under a "philosophy of more employment" will produce spectacular results, notably the young unemployed graduates migrating to three urban towns in the state.

A review of the validity of the model shows that 30 years or more may elapse between the conceptions of the needs and full utilization of the model. In addition, most component of the model involves very large investment cost. The sunk costs of the project completion are very important and make the corresponding decisions economically rigid and relatively irreversible.

7.5 CONCLUSION

The optimization system is an optimal solution and a feasible solution. The most favorable value of the objective function is the largest value for maximum environmental benefit (Benefit cost Ratio) and smallest value for a minimization problem of cost of maintenance.

This model presented here illustrates mix basic solution integrated planning of state and some Local Government Solid Waste Management System in Anambra State. The optimization model was developed with the objective of allowing the Board of ANSEPA to capture practically all aspect of waste management and it is planning problem (All integrated into, personnel sundry, equipment maintenance, and purchase). It contains many innovative features and removes many limitation frequently encountered in often existing optimization modeling for waste management.

Moreover financial constraint causes delay in the models effectiveness and efficiency. As in the Nigeria factor, financial resources are usually difficult to access, in that model goals can be delayed overhead waste collection, disposal and planning management, and a whole lot could be disrupted. When the complexity of solid waste management planning increases, system engineering tools can assist municipal and local decision makers in handling the complex planning situation.

KEYWORDS

- **Deterministic model**
- **Infectious disease**
- **Integrated solid waste management**
- **Maximum benefit**
- **Minimum cost**
- **Morbidity and mortality**

REFERENCES

Gottinger, H. W.. *Economic Model and Application of solid Waste Management.* Gordon and Breach Science Publisher, New York, 1991 (1986).

Kaila, J. *Mathematic Model for Strategic Evaluation of Municipal Solid Waste management system.* technical research centre of Finland Publication 40, Espoo. Finland (1987).

Mackenzie, L. D.and David, A. C.. Introduction to Environmental Engineering Mc Graw Hill series, pp. 630–701 (1998).

Peavy, H. S., Rowe D. R., and Techobanoglous, G. Environmental Engineering McGraw Series, pp. 594–652 (1985).

Sundberg, J. A System Approach to Municipal Solid Waste Management. A Pilot Study of Goleboy. *Waste Manage. Res.,***12**, 7 (1993).

Tchobanoglous, G., Theisen, H., and Vigil, S.. Integrated Solid Waste Management McGrow Hill Series, p. 7 (1993).

Tomas, F., Jared, C., and David, M.. The Mathematical Programming Sequencing Model, pp. 112–128 (1979).

8 Optimizing Urban Material Flows and Waste Streams in Urban Development Through Principles of Zero Waste and Sustainable Consumption

Steffen Lehmann

CONTENTS

8.1   INTRODUCTION

Beyond energy efficiency, there are now urgent challenges around the supply of resources, materials, energy, food and water. After debating energy efficiency for the last
decade, the focus has shifted to include further resources and material efficiency. In
this context, urban farming has emerged as a valid urban design strategy, where food
is produced and consumed locally within city boundaries, turning disused sites and
underutilized public space into productive urban landscapes and community gardens.
Furthermore, such agricultural activities allow for effective composting of organic
waste, returning nutrients to the soil and improving biodiversity in the urban environment. Urban farming and resource recovery will help to feed the 9 billion by 2050
(predicted population growth, UN-Habitat forecast 2009). This chapter reports on
best practice of urban design principles in regard to materials flow, material recovery,
adaptive re-use of entire building elements and components ("design for disassembly"; prefabrication of modular building components), and other relevant strategies to
implement zero waste by avoiding waste creation, reducing wasteful consumption and
changing behavior in the design and construction sectors. The chapter touches on two
important issues in regard to the rapid depletion of the world's natural resources: the
built environment and the education of architects and designers (both topics of further
research). The construction and demolition (C&D) sector: Prefabricated multi-story
buildings for inner city living can set new benchmarks for minimizing construction
wastage and for modular onsite assembly. Today, the C&D sector is one of the main
producers of waste; it does not engage enough with waste minimization, waste avoidance and recycling. Education and research: It is still unclear how best to introduce a
holistic understanding of these challenges and to better teach practical and affordable
solutions to architects, urban designers, industrial designers, and so on. How must urban development and construction change and evolve to automatically embed sustainability in the way we design, build, operate, maintain and renew/recycle cities? One
of the findings of this chapter is that embedding zero waste requires strong industry
leadership, new policies and effective education curricula, as well as raising awareness (through research and education) and refocusing research agendas to bring about
attitudinal change and the reduction of wasteful consumption.

Since, the industrial revolution, mankind has constantly expanded and increased
industrial production and urbanization, using massive resources of materials and energy. The mass consumption of resources raises serious problems such as global warming, material depletion and enormous waste generation.

This chapter explores the notion of sustainable urban metabolism and "zero waste". There is now a growing interest in understanding the complex interactions and feedbacks between urbanization, material consumption and the depletion of our resources. The link between increasing urbanization and the increase of waste generation has been established for some time. However, the impact of urban form and density on resource consumption is still not fully understood. Human population on the planet has increased 4-fold over the last 100 years, while—in the same time period—material and energy use has increased 10-fold (World Urbanization Prospects, 2010). The United Nations forecast that the world's urban population will increase by 2.7 billion people between 2010 and 2050. But how can urbanization of our planet continue with such devastating effects?

Based on our wasteful patterns of urban development, it's time to rethink development practice and urban form (Satterthwaite, 2009). However, to formulate better urban responses requires a full awareness of the impacts and reasons for current global change, which mainly occurs through:

- Demographical changes
- Growing social disparities
- Continuing urbanization processes with rapidly expanding cities
- Growing demand for resources (materials, energy, water)
- Loss of biodiversity and habitat, and
- Continuing production methods of industry and agriculture often too material and energy intensive and therefore unsustainable.

The pace of urbanization is increasing and cities face new challenges from the effects of human activity on global systems, which in turn impact on urban life. Climate change is a significant one of those challenges. It is apparent that cities are the main consumers of materials, energy, water, and food, and hence they are the main sources of greenhouse gas (GHG) emissions associated with climate change. Holistic understanding and integrated approaches to design, planning and urban management are essential to effective resolution of urban problems. In most countries, cities keep expanding with growing populations. It is particularly important to include the peri-urban areas and suburbs in any research and analysis, as they represent the areas of interaction between the urban and rural contexts, where fertile agricultural land and precious landscape is gradually lost as a food source.

Beyond energy efficiency, there are now urgent challenges around the supply of resources, materials, food and water, and after debating energy efficiency for the last two decades, the focus has shifted to include resource and material efficiency. Waste was once seen as a burden on our industries and communities; however, shifting attitudes and better understanding of global warming and the depletion of resources have led to the identification of waste as a valuable resource that demands responsible solutions for collecting, separating, nurturing, managing and recovering. In particular, over the last decade, the holistic concept of a "zero waste" life-cycle has emerged as a cultural shift, as a new way of thinking about the age-old problem of waste and the economic obsession with endless growth and consumption.

Emerging complex global issues, such as health and the environment, or lifestyles and consumption, require approaches that transcend the traditional boundaries between disciplines. The relationship between efficiency and effectiveness is not always clear: high efficiency is not equal to high effectiveness, while recovery offers another side of those two notions. Today, it is increasingly understood that the same way we discuss energy efficiency; we need also to discuss resource effectiveness and resource recovery. This includes waste minimization strategies and the concept of "designing waste out of processes and products" (as mentioned, for instance, in (South Australia's Draft Waste Strategy, 2010)).

Every municipality or company can take immediate action to identify its own particular solutions. Separating recyclable materials, such as paper, metals, plastics and glass bottles, and consolidating all identified waste categories into one collection point, are some basic measures. However, a waste stream analysis will have to be conducted at an early stage, which will involve taking an inventory of the entire waste composition, measuring the volumes of different material categories and its origin and destination. A database will then need to be created to enable the municipality to track all waste types and to cross reference by facility type, so the amount and type of waste each facility, district or precinct generates can be identified, thus pinpointing where reductions can occur.

For centuries, waste was regarded as "pollution" that had to be hidden and buried as landfill. Today, the concept of "zero waste" directly challenges the common assumption that waste is unavoidable and has no value by focusing on waste as a "misallocated resource"[1] that has to be recovered. It also focuses on the avoidance of waste creation in the first place (e.g., reducing construction waste). That we are a wasteful nation is illustrated by the fact that over 40% of our daily food is thrown out and wasted (Environmental Protection Agency, 2009). Recent research found that family size and household income are primary determinants of household waste, while the affect of environmental awareness on waste generation behavior is surprisingly small.

This, of course, raises much wider social questions of attitude and behavior, and our wastefulness has further implications on future urban development. How will we design, build, operate, maintain and renew cities in the future? What role will materials play in the "city of tomorrow"? How can we increase our focus on more effective environmental education for waste avoidance? And how we will need to better engage sustainable urban development principles and zero waste thinking? These are some of the topics discussed in this chapter.

8.2 THE LINK BETWEEN WASTE AND URBANIZATION

8.2.1 Limits of Growth: Understanding Waste as a Resource and Part of a Closed-Cycle Urban Ecology

In recent years, the need for more sustainable living choices and a focus on behavioral change has increasingly been articulated. The estimated world waste production is now around 4 billion tons of waste per annum, of which only 20% is currently recovered or recycled. Globally, waste management has emerged as a huge challenge, and it is time that we took a fresh look at how we can best manage the waste and material streams of cities and urban development. The issue of our city's ever growing waste production is of particular significance if we comprehend the city as a living ecosystem with closed loop management cycles (see Figures 8.1 and 2).

There are some serious implications around the topic of waste. It is obvious that it is not just about waste recycling, but also waste prevention, following the waste hierarchy diagram (see Figure 8.3). We must give prevention more priority, as the saying goes: "An ounce of prevention is worth a pound of recycling." Avoidance is the priority, followed by recycling and "waste engineering" (up-scaling), to minimize the amount that goes to waste incineration.

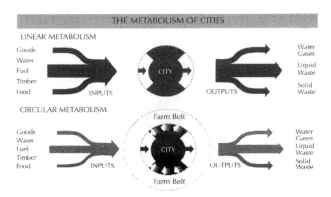

FIGURE 8.1   The flow of natural resources into cities and the waste produced (recovering waste streams) represents one of the largest challenges to urban sustainability. Circular, looping metabolisms are more sustainable, compared to linear ones. This also has economic advantages. Recycling will continue to be an essential part of responsible materials management, and the greater the shift from a "river" economy (linear throughput of materials), towards a "lake" economy (stock of continuously circulating materials), the greater are both the material gains and GHG reductions (Diagram source: (Girardet, 1999, 2008), republished in (Rogers and Power, 2000)).

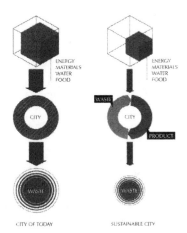

FIGURE 8.2   Diagram illustrating the input and output of cities, comparing the "conventional" city with the more sustainable city on the right (Diagram source: (Blue, 2010)).

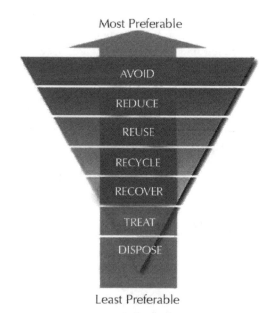

FIGURE 8.3 The waste hierarchy diagram illustrates how waste avoidance is preferred, above re-use and recycling. Disposal in landfill represents the lowest level of the waste hierarchy (diagram: courtesy of the author, 2010). On the municipality level, a more strategic charging structure (levies) for waste disposal can accelerate sustainable waste management and reward residents who are separating their waste.

A particular concern is disposal of electrical and electronic equipment, known as *e-waste*. Of about 16.8 million televisions and computers that reached the end of their useful life in Australia in 2008 and 2009, only about 10% were recycled. Most of the highly toxic e-waste still goes into landfills, threatening ground water and soil quality, and an unknown proportion is shipped overseas (legally and illegally), mainly to China, leading to major environmental problems in these importing countries. About 37 million computers, 17 million televisions, and 56 million mobile phones have already been buried in landfills around Australia. This waste contains high levels of mercury and other toxic materials common to electronic goods, such as lead, arsenic, and bromide. Several countries are actively pushing for industry-led schemes for collecting and recycling televisions, printers, and computers, known as extended producer responsibility (EPR) and product stewardship. In addition, we must expect that the amount of e-waste created in the developing world will dramatically increase over the next decade (Easton, 2010).

Discharges are a threat to soil and groundwater, and methane gas discharges (mainly from organic waste in landfill) are a threat in the atmosphere. In the meantime, many large cities are producing astronomical amounts of waste daily and are running out of landfill space. Incineration of waste has gone out of fashion, as it has the disadvantage that it releases poisonous substances, such as dioxins and toxic ash, into the environment. Burning waste with very high embodied energy is generally not an efficient way of dealing with resources. Environmental groups have successfully prevented the construction of new waste incinerators around the world. Such linear systems (e.g., burning waste) have to be replaced with circular systems, taking nature as its model. Much more appropriate is a combination of recycling and composting. Today, recycling 50–60% of all waste has become an achievable standard figure for many cities (e.g., the Brazilian model city of Curitiba has managed to recycle over 70% of its waste since 2000; and the city of San Francisco has arrived at 77% diversion rate from landfill in 2010).

Organic waste is playing an increasingly important role. The small Austrian town of Guessing, for instance, activates the biomass from its agricultural waste and has reached energy autonomy by composting and using the bio-energy to generate its power. In the available literature, a recommended split for a city can be found, where no waste goes to landfill:

- Recycling and reusing min   50–60%
- Composting of organic waste   20–30%
- Incineration of residual waste (waste-to-energy)   max. 20%

Steel is by far the most recycled material worldwide (it has the longest "residence time"). However, recent research from Veolia Research Group (Grosse, 2010) shows that recycling in itself is inefficient in solving the problem, as it does not deliver the necessary "decoupling" of economic development from the depletion of non-renewable raw materials. Grosse and others argue that "the depletion of the natural resource of raw material is inevitable when its global consumption by the economy grows by more than 1% per annum. The only effect of recycling is that the curve is delayed." There is evidence that recycling can only delay the depletion of virgin raw materials for a few decades at best. Research shows that only recycling rates above 80% would allow a significant slowdown of the depletion of natural resources. This means that the actual role of recycling to protect resources is not significant for non-renewable resources whose consumption tends to grow above 1% per year.

Even though it is an important component, sustainable development policies cannot rely solely on recycling. Policies need to aim at reducing the consumption of each non-renewable raw material so that the annual growth rate remains under 1%. Decoupling economic development from materiality seems to be the only long term solution. Recycling is not so much the primary goal. The objective is not so much to reduce the amount of waste in general, but, rather, to encourage a reduction in the quantities of materials used to make the products that will later become waste. *Waste is nutrients. Waste is precious. We should learn from Nature: Nature does not know "waste". In Nature, one species' waste is another species' resource* [2,3] (Braungart and McDonough, 2002).

8.2.2   Zero Waste and Closed Loop Thinking in the Construction Sector

There is a growing interest from architects in zero waste concepts. Cities and urban development are the areas where all concepts come together and can be embedded into practice, into redesigning urban systems with zero waste and material flow in mind, by transforming the existing city and upgrading its recycling infrastructure in low-to-no carbon city districts. It's timely to rethink prefabrication and "design for disassembly" building resilience into urban systems. This will change the way we design, build and operate city districts in future (acknowledging that zero waste is much wider and complicated than expected at the first glance, and that we still have long distance from zero emission to zero waste in regard to the construction sector). For instance, façade systems made of composite materials create recycling and resource recovery problems. No debris should go to landfill. Concrete companies should use sustainable, recycled aggregates (RAs). Concrete was previously regarded as being difficult to be recycled, as closed loop recycling for concrete structures is expensive. But concrete-related waste is now increasingly used as RA for new concrete structures, and intensive research is carried out in Japan and China on new concrete recycling methods.

Urban planners frequently raise the question about which is the best scale to operate on for introducing "zero waste". The city district as a unit appears to be a good, effective scale. It means rejoining the urban with the rural community; therefore neighborhood and precinct planning must consider the climate crisis. For instance, planning better cities requires that composting facilities and recycling centers are in close proximity to avoid transporting materials over long distances. Reducing energy embodied in construction materials is an important strategy for mitigating our fossil-fuel dependency. Keeping the existing building stock is important, as the most sustainable building is always the one that already exists. Retrofitting existing districts is, therefore, essential.

8.2.3 Constantly Growing Amounts of Waste—What Can Be Done?

Global population growth is expected to stabilize in 2050 at around 9 billion human beings (World Population Growth Forecast, 2010). However, population growth is far from being the main driver of recent economic expansion and the increase of consumption of materials, water, fossil-fuels and resources. The process by which emerging countries catch up with the standard of living of more advanced economies is, in fact, an even more powerful actuator.

As a consequence of this "catching up", waste is accumulating in the oceans. In recent years, our oceans have devolved into vast garbage dumps. Thousands of tons of waste are thrown into the sea each year, endangering humans and wildlife. Since, the world's oceans are so massive, few people seem to have a problem with dumping waste into them. However, most plastics degrade at a very slow rate, and huge amounts of them are sloshing around in our oceans. Wildlife consumes small pieces, causing many of them to die as the plastics are full of poisons. Some plastic products take up to 200 years to degrade. Every year, around 250 million tons of plastic products are produced, and much of this produce ends up in the oceans. The "Great Pacific Garbage Patch" is half the size of Europe, and in the Atlantic huge amounts of plastic garbage have recently been discovered (Waste Report, 2010); the highest concentration being found close to Caribbean islands, with over 2,00,000 plastic pieces per sq km. In the North and Baltic seas, although dumping in them has been illegal for over two decades, the amount of waste found in them has not improved. It is estimated that each year 20,000 tons of waste finds its way into the North Sea, primarily from ships and the fishing industry (World Urbanization Prospects, 2010). Experts warn that we have reached a point where it's becoming dangerous for humans to consume seafood. A big problem is the throw-away plastic water bottles made of PET, not only because they significantly contribute to waste creation and CO_2 emissions from transporting drinking water around the globe, but they also release chemicals suspected of being harmful to humans into the water. Together with the largest oil spill in human history, the devastating oil spill in the Gulf of Mexico (2010), it shows how advanced humanity's destruction of entire ecosystems in the oceans has become.

Given these conditions, the international community has been pushing for four decades for massive bureaucratic efforts aimed at clearing the oceans of waste. In 1973, the United Nations sponsored a pact for protecting the oceans from dumping, and in 2001 the European Union established directives that forbade any dumping of maritime waste into the ocean while in port. However, such directives have been ineffective and many experts agree that laws and international efforts aimed at protecting the oceans have failed across the board.

Today, no other sector of industry uses more materials, produces more waste and contributes less to recycling than the construction sector (Environmental Protection Agency, 2009).

With the constant increase in the world's economic activity, there has been a large increase in the amount of solid waste produced per head of population. The waste mix (industrial and urban) has become ever more complex, often containing large amounts of toxic chemicals. Obviously, the first aim of a sustainable future is to avoid the creation of waste and to select materials and products based on their embodied energy, on their life-cycle assessment and supply chain analysis. This needs to be understood holistically. Transportation of input materials, as well as the transportation of the final product to consumers (or to the construction site), is a common contributor to GHG emissions. The way in which a product uses resources such as water and electricity influences its environmental impact, while its durability determines how soon it must enter the waste stream. Care needs to be taken in the original selection of input materials, and that the type of assembly used influences end-of-life disposal options, such as ease of recyclability or take-back by the manufacturer. With a huge amount of waste still going to landfill, drastic action is required in urban planning to develop intelligent circular metabolisms for districts, and waste collection and treatment systems that will eliminate the need for landfills. Even so, recycling is only halfway up the waste hierarchy, the greenhouse gains lying in the upper half (waste avoidance and reduction) are, largely, yet to be tapped. The focus of attention needs now to expand from the downstream of the materials cycle, from a post-consumer stage, to include the upstream, pre-consumer stage, and behavioral change (see diagram Figure 8.1).

Figure 8.1 illustrates the concept of a *circular* (looping) urban metabolism: the current production-consumption system is typically *linear* (as in a pipeline) and extends from manufacturing through use to end of life, followed by either recycling or landfill. The idea that this system must be reconfigured in order to promote a series of closed loops whereby all material and products are re-used or recovered is not new and has been raised many times, however, has not been adapted by the construction sector. Figure 8.4 illustrates how future buildings will produce energy and even food.

While the worldwide international average for daily waste generation is about 11.5 kg per capita, countries like Kuwait and United Arab Emirates top the list, generating an average of over 3.5 kg of waste per person per day (in comparison, the average Australian resident dumps 1.1 tons of solid waste per year, this is also around 3 kg per day). According to the "polluter pays" principle, policies penalize those who generate large amounts of waste. Collecting, sorting and treating waste incurs huge costs, so the focus has to be on avoiding and minimizing waste creation in the first place, in the office, in industry, in households. Waste-wood-to-energy has frequently become an important component of energy concepts for city districts. Waste management and recycling schemes have greatly reduced the volume of waste being "landfilled". Waste segregation and recycling has also substantial economic benefits and creates new jobs.

Re-using building components and integrating existing buildings (instead of demolition) is a basic principle of any eco-city and eco-building project (Lehmann, 2008).

FIGURE 8.4 Urban farming, designing the "Carrot City": with finite cropland to feed a growing global population, concepts are now being developed that will build vertical farms, where buildings' roofs and facades become sites for urban agriculture. Rotating hydroponic-farming systems give the plants the precise amount of light and nutrients they need, while vertical stacking enables the use of far less water than conventional farming (Project illustration: (Carrot City, 2009)).

8.2.4   Changing Manufacturing and Packaging Processes towards Life-Cycle Oriented Practices

New agreements with industry have to be made to dramatically reduce waste from packaging. On the way towards a zero waste economy, manufacturers will increasingly be made responsible for the entire life-cycle of their products, including their recyclability, by introducing an EPR policy. Luckily, many companies are now doing extraordinary things in the area of recycling and are prolonging the life-cycle of products. For instance, Ohio-based firm *Weisenbach Recycled Products*, a manufacturer of consumer goods made from recycled materials holds numerous patents on recycling, awareness and pollution prevention products. It is both a specialty printing firm and an innovative recycler of waste and scrap, repurposing and "up-cycling" such materials as plastic caps, glass bottles, and circuit boards into over 600 promotional items and retail consumer products. According to the company's president, Dan Weisenbach, there has been a changing perception in the business world, where you are more valued if your company is a "certified green business": "Even though conservation has been a core principle in our culture since we started, we believe it is important that we take a step to formalize our commitment to sustainable business. The competitive landscape has shifted and it is important for a company to have a history of environmental leadership and integrity. Choosing to voluntarily document all our efforts in an annual sustainability report is a demonstration of this commitment. We have moved past the *bigger is better* era. People want to do business with companies they can relate to and who share their values" (Weisenbach, 2010).

For centuries, waste was regarded as pollution that had to be collected, hidden and buried. Today, waste is no longer seen as something to be disposed of, but as a resource to be recycled and reused. It is clear that we need to close the material cycle loop by transforming waste into a material resource. Over the next decades, the Earth will be increasingly under pressure from population growth, continuing urbanization and shortage of food, water, resources and materials. Waste management, optimizing waste streams and material flows are some of the major challenges concerning sustainable urban development. There is a growing consensus that waste should be regarded as a "valuable resource and as nutrition"[4] (Braungart and McDonough, 2002). It has been argued that the concept of "waste" should be substituted by the concept of "resource". McDonough and Braungart point out that the practice of dumping waste into landfill is a sign of a "failure to design recyclable, sustainable products and processes." All eco-cities have to embed zero waste concepts as part of their holistic, circular approach to material flows (see diagram Figure 8.2).

Design for Disassembly means the possibility of reusing entire building components in another future project, possibly 20 or 30 years after construction. It means deliberately enabling "chains of reuse" in the design, and to use light-weight structures with less embodied energy, employing modular prefabrication. Recycling resources that have already entered the human economy uses much less energy than does mining and manufacturing virgin materials from scratch. For instance, there is a 95% energy saving when using secondary (recycled) aluminum; 85% for copper; 80% for plastics; 74% for steel; and 64% for paper (Fischer-Kowalski, 1998). Through re-use and recycling, the energy embodied in waste products is retained, thereby slowing down the potential for climate change. If burned in incinerators, this embodied energy would be lost forever. It becomes obvious that all future eco-cities will have to integrate existing structures and buildings for adaptive re-use into their master planning. Based on life-cycle assessment, the most sustainable building is likely to be the one that already exists.

In closed loop systems, a high proportion of energy and materials will need to be provided from re-used waste, and water from wastewater. We can now move the focus to waste avoidance, behavioral change and waste reduction.

8.2.5   A Closed-cycle Urban Economy will deliver a Series of Further Advantages

- It avoids waste being generated in the first place (and therefore reduces CO_2 emissions).
- It creates closed loop eco-economies and urban eco-systems with green collar jobs.
- It helps transform industries towards a better use of resources and non-polluting (non-toxic), cleaner production processes, and extend producer responsibility.
- It delivers economic benefits through more efficient use of resources.
- It supports research into durable, local goods and products that encourage reuse.
- It advocates green purchasing and a product stewardship framework.

The EPR places the responsibility of the future of an item of waste on the initial producer of that product (instead of on the last owner, as in traditional segmentation) (Linacre, 2007; ZWSA Survey, 2009). This leads to the practice whereby an increasing number of manufacturers include in the sale of goods a service for the future recovery and the processing of the product at the end of its useful life.

It also includes extending the responsibilities to consumers to participate in re-cycling schemes. A recent survey showed that 83% of Australians wanted a national ban on non-biodegradable plastic bags, while 79% wanted electronic waste (e-waste) to be legally barred from landfills[5]. Cities will always be a place of waste production, but there are possibilities available that will help them achieve zero waste, where the waste is either recycled, reused or composted (using organic waste for biomass). The Masdar-city project in the UAE is a good example of a zero waste city, as is the large Japanese city of Yokohama, which reduced its waste by 39% between 2001 and 2007, despite the city growing by 1,65,000 people during this period. They reached their goal by raising public awareness about wasteful consumption and through the active participation of citizens and businesses. In Australia, the *Zero Waste SA* initiative by the South Australian government is highly commendable (Meadows et. al., 1972).

8.2.6   Behavior Change for Waste Prevention

The growth of the economy cannot continue endlessly (a fact already pointed out by (On sustainable urban development, 1987)). Our increasing affluence allows us to accumulate massive amounts of stuff, and we build increasingly larger dwellings to store it. So the core question is about how to best change behavior and shift attitudes to reduce consumption (and therefore avoiding the creation of waste in the first place). How do we convince society to consume less? Education programs aimed at all levels of schooling has proven to be effective. Public education aimed at "zero waste" participation is surely a key to success. Changing behavior is easier in smaller towns, but is more difficult in large cities. As has already been pointed out, education to raise awareness is essential, but equally important is that the rules of waste separation are well explained. This suggests that the real problem is not technology, but acceptance and behavior change. What is needed is social innovation rather than a sole focus on technological innovation. The necessary connection between waste policies and emission reductions are not always well understood and made.

So, what are the main barriers to zero waste?

- Short term thinking of producers and consumers
- Lack of consistency in legislation across the states
- Procurement *vs.* sustainability: the attitude that the cheapest offer gets commissioned
- Lack of community willingness to pay

The increase in world flows of scrap, e-waste, recovered plastics and fibers has turned developed countries into a source of material supply for informal trade in emerging countries (United Nations Human Settlements Programme, 2010).

8.2.7   Introducing Product Stewardship: Consumptive Lifestyle Decisions and Household Practices

There is a clear need for designers to focus more attention on the throughput of material goods consumed in our everyday life rather than just end use energy consumption (Hobson, 2003; Lane et. al., 2009; Tonkinwise, 2005). *Product stewardship* refers to the responsible management of manufactured goods and materials. On the production side, product and industrial designers are critical for stewardship models that go beyond materials recycling (e.g., extended producer responsibility), however, until now design issues have not figured strongly in product stewardship schemes, and there is not enough attention to product stewardship of new goods and their disposal at the end of use.

Drawing on social practice theory[6,7,8,9] (Barr and Gilg, 2006; Schatzki, 2003) consumption within the household can be explained as the outcome of the relations between household routines and surrounding material systems of provision. For product/industrial designers, social practice is a relatively new area of study, and practice-orientated design is only slowly moving beyond the tradition of designers (just focusing on products in isolation)—instead acknowledging that—material artifacts themselves configure the needs and practices of those who use them (Shove et. al., 2007).

However, achieving net reduction in material and energy flows implies changes in design and household practices, and the introduction of product stewardship models. Current household practices around the acquisition, use and disposal of common household furnishings and electronic goods depend largely on household type and urban context, including house size and location to public transport. Dey et al. and Perkins et al. note that household practices and consumption vary across households and their urban contexts (e.g., suburban dwelling versus inner-city apartment) (Dey et. al., 2007; Perkins et. al., 2009). Products themselves place constraints on how householders may exercise stewardship responsibilities, which indicates that household decisions concerning product stewardship, acquisition and divestment are mainly influenced by a range of factors including the physical spaces of the home, issues of wealth and social status (life stage), cultural values and habits established over time. There is still a need for more research in the question, how can product stewardship be extended through new product design in order to explicitly include household consumers' acquisition and better use of products, as well as end of life disposal options.

8.3   CASE STUDIES OF WASTE MANAGEMENT

The following case studies include details of how some cities and regions are trying to overcome the barriers to achieving "zero waste". The cases are looking at waste stream management in the developed world (Australia and Denmark) and at two large cities in the developing world (Delhi and Cairo, both rapidly expanding cities).

Case 1: South Australia's leadership in waste management and resource recovery
South Australia, over the last 5 years, has produced a document on zero waste principles, the "Draft South Australia's Waste Strategy 2010–2015" (South Australia's Draft Waste Strategy, 2010). The strategy offers strong guidelines for SAs waste recycling and waste avoidance efforts, and has a 5 year timeframe. The strategy's focus is on two objectives: "Firstly, the strategy seeks to maximize the value of our resources; and secondly, it seeks to avoid and reduce waste." These two objectives are inter-related, and some actions apply to both objectives, proposing new targets for municipal, commercial and industrial and C&D waste streams. Zero Waste SA is one of the few zero waste government agencies in the world and is at the forefront of waste avoidance in Australia. Zero Waste SA was established in 2003 and is financed by government levies from landfill. The agency pioneered the introduction of the ban on checkout style plastic bags in Australia, in May 2009, and formulated the campaign slogan: "I recycle correctly and everyone wins".

To be able to increase recycling and to reduce consumption, we need to fully understand the composition of household waste. Only by separation at the source (point of waste creation), can we reach high recycling rates. Interestingly, recent research at the UniSA indicates that the composition of waste varies according to the income level of the people producing the waste. For instance, the amount of food waste tends to be greatest among lower-income earners (this is because as income increases there is generally less food waste as consumers purchase greater amounts of prepared food relative to fresh food).

The SA Draft Waste Strategy policy is no unique case or exemption. All of the European Union member states must compile a waste prevention program by the end of 2013, as required by the 2008 revision of the "Waste Framework Directive". The EU guidelines are intended to support the formulation of such programs based on 30 best practices identified by the European Commission.

Case 2: The waste situation in New South Wales (NSW), Australia: a looming crisis?
Australia is the third highest generator of waste per capita in the developed world. In July, 2006, only around 50% of waste collected in the state of NSW was recycled. Of course, it's always cheaper to simply bury waste than to treat it, but that has dangerous side effects. For instance, electronic waste is still filling up Australian and US landfills (something not allowed in the EU for 10 years), contaminating soil and groundwater with toxic heavy metals. In the meantime, a waste crisis is looming: the City of Sydney's four landfill sites (Eastern Creek, Belrose, Jacks Gully and Lucas Heights) are reaching capacity and will be full by 2015, according to a recent independent *Public Review Landfill Capacity and Demand Report* (State Government of New South Wales, 2009). The city's annual 2 million tons of waste will have to be moved 250 km south, by rail, to Tarago. For a long time, the state government has been inactive and

has failed to make the recycling shift. It lacks recycling policies and investment in re-cycling technology. Recycling needs to be made cheaper than land filling, and strong economic incentives are required, as are strategies to get households to dramatically reduce the creation of waste (for instance, by reducing bin sizes, raising awareness and by introducing the three-bin system to separate organic/garden waste, recycling, and residual waste).

The situation in the UK is similar. Mal Williams, CEO of Cylch (a major recycling company in Wales, UK), points out that "90% of household waste is actually reusable without the need for incineration. Waste means inefficiency and lost profit for all" (Williams, 2010).

While Sydney's landfill sites are rapidly filling up, and the NSW government has currently no clear plan to address the crisis, Sydney's waste is forecast to keep growing by at least 1.4% a year (due to population increase and increasing consumption). Curbside recycling collected in NSW increased from 4,50,000 tons in 2,000 to 6,90,000 tons in 2007. To make things worse, the NSW government rose over $260 million in waste levies but returned just 15% ($40 million) of that to local councils for recycling initiatives (Public Review Landfill Capacity and Demand Report, 2009). By contrast, the state government of Victoria gives better support: it raised $43 million in landfill levies and gave it straight back to the agencies responsible for waste management. Despite the smaller levy, Victoria recycled almost 20% more waste than NSW in 2009. The federal government will introduce a *National Waste Policy* in 2011 (aiming for a 66% landfill reduction by 2014) and hopes are high that this will bring about the urgently required changes.

Case 3: Waste management case study from Aalborg, Denmark
Developed countries such as Germany, Japan and Denmark are worldwide leaders in waste management. For instance, in some Japanese municipalities up to 24 different categories of waste are separated.

It is timely that we better integrate the linkages between material flow, use and recovery with energy and water consumption. To date, little research has been done on measuring the impact of waste treatment systems themselves and waste management changes over the longer term. For instance, the Danish city of Aalborg has proven that better waste management can reduce GHG emissions and that a municipality can produce significant amounts of energy with sustainable waste-to-energy concepts. Two Danish researchers, Poulsen and Hansen, used historical data from the municipality of Aalborg to gain a longer-term overview of how a "joined-up" approach to waste can impact on a city's CO_2 emissions. Their assessment included sewage sludge, food waste, yard waste and other organic waste. In 1970 Aalborg's municipal organic waste management system showed net GHG emissions by methane from landfill of almost 100% of the total emissions. Between 1970 and 2005, the city changed its waste treatment strategy to include yard waste composting, and the city's remaining organic waste was incinerated for combined-heat and-power (CHP) production. Of this, waste incineration contributed 80% to net energy production and GHG turnover, wastewater treatment (including sludge digestion) contributed another 10%, while other waste treatment processes (such as composting, transport, and land application of treated waste) had minor impacts. "Generally, incineration with or without energy production, and biogas production with energy extraction, are the two most important processes for the overall energy balance. This is mainly due to the substitution of fossil fuel-based energy," says Poulsen. The researchers calculated that the energy potential tied up in municipal organic waste in Denmark is equivalent to 5% of the country's total energy consumption, including transport. They also predicted that further improvements by 2020 were possible, by reducing energy consumed by wastewater treatment (for aeration), increasing anaerobic digestion, improving incineration process efficiency and source separating food waste for anaerobic co-digestion.

Understanding of natural systems, this is a pioneering demonstration on how technology can be harnessed to resolve environmental challenges. Aalborg's progress shows how far-reaching waste management can be in attaining energy and GHG reduction goals, and should offer encouragement to other cities embarking on greener waste management strategies for the future (On Poulsen and Hansen's Research, 2009; Poulsen and Hansen, 2009). The potential for emission reduction in waste management is very big. It is estimated that within the European Union, municipal waste management reduced GHG emissions from 64 to 28 million tons of CO_2 per year between 1990 and 2007, equivalent to a reduction from 130 to 60 kg CO_2 each year per capita. With such innovation in waste treatment, the EU municipal waste sector will achieve 18% of the reduction target set for Europe by the Kyoto agreement, before 2012.

8.4 SCARCITY OF RAW MATERIALS, METALS, RESOURCES

8.4.1 Using Fewer Materials to Better Exploit the Value of Waste

Energy cost is not limited to heating or cooling energy or lighting energy; it is also related to all material flows relevant to buildings. For instance, waste from the production of construction materials and components can be much greater than all other waste streams. To make it easier for architects and planners to specify materials according to their impact (including impacts caused by material extraction, or waste creation from the production process), information on materials and components needs to be readily available. Different from the *Club of Rome's* warning of 1971, today, the "limits of growth" are defined by climate change and the depletion of material resources. We see an increasing challenge through the scarcity of raw materials, especially metals such as lead, copper and zinc. With natural resources and materials about to run out, we need better resource protection and more effective ways to use them. Several essential metals and resources are already becoming less available, for example most platinum, zinc, tantalum, lead, copper, cadmium, wolfram and silicon is concentrated in the hands of three countries, under the control of three large companies. This will soon create major challenges for industries in Europe and the US that use many of these metals in their manufacturing (such as televisions or computers). In a resource-constrained future we will see more:

- Recycling-friendly designs, with EPR
- Multiple-use (multi-function) devices and expanded product lifecycles,
- Long-life products and buildings, with optimized material use,
- Products using less packaging,

- A variety of ways to avoid the loss of resources during the product's life-cycle,
- Resource recovery through forward thinking reuse, remanufacturing and recycling.

Waste that contains precious minerals, rare earth, metals and other nutrients is now understood to be valuable, and organic waste must be returned to the soil. The survival path and rebound effect of materials is understood as extremely critical. Will our landfill sites of today become the "urban mines" of the future? We can observe the emergence of a new sustainable industrial society, where new industrial systems are introduced that better reuse and recycle waste, and which are based on a new circular flow economy (Faulstich, 2010; Girardet, 2010). In the meantime, the depletion of several natural deposits is drawing closer. In 2008, the *Institut der Deutschen Wirtschaft* (IDW) estimated the availability and coverage of essential resources and selected metals, as part of a risk assessment for the German industry in response to the threat caused by scarcity of raw materials[10,11]. It found:

Lead	20 years reserves available, estimated
Zinc	22 years
Tantalum	29 years
Copper	31 years
Cadmium	34 years
Wolfram	39 years
Nicke	144 years

These metals are becoming scarce and consequently more expensive, for example iron ore, lithium and copper are already much rarer than oil. In addition, it is also important to know what kinds of products we buy. For instance, 40% of the products in our weekly shopping basket contain palm oil, which, if not produced sustainably, can cause deforestation of ecologically precious rainforests. A more conscious use of materials, metals, resources and products is an imperative, supported by reuse and recycling.

Cities are resource-intensive systems. By 2030, we will need to produce 50% more energy and 30% more food on less land, with less water and fewer pesticides, using less material (Head, 2009).

8.4.2 The Need for Changing the Practice of Packaging with a "Product Stewardship" Program

There is a growing need for use of truly compostable packaging, where everything that arrives at the consumer is useful and does not create waste.

In future, (with EPR) the user of packaging will have to pay for the collection of that packaging (Easton, 2010). The rising costs of waste from landfill levies will become its main driver. Essentially, one needs to ask: How much packaging is really necessary? Can the product be packed in another way? There is a need for leadership from a select group of companies (this is usually not more than 5% of all companies) to show how packaging can be reduced, or how products can be taken back from the consumer once the end of life-cycle has been reached, as is done with old tyres. Ikea and Woolworth have been setting new standards in this area, and BASF only puts new products on the market when there is evidence that the new product has a better life-cycle assessment than the previous one. There have been innovative recycling initiatives for mattresses, bicycles, carpets, paints, construction timber and furniture. We will need more products to be manufactured differently to how they are made now, with zero waste concepts in mind and also taking the EPR principle seriously. In the US, 44% of all GHG emissions result from transporting and packaging products, illustrating the large potential in this field.

8.5   A LACK OF WASTE MANAGEMENT FRAMEWORKS IN THE DEVELOPING WORLD

8.5.1   Informal Waste Recycling Sectors in the Developing World

A staggering 95% of global growth over the next 40 years will happen in Asia, Africa, Latin America and the Caribbean, according to the Population Reference Bureau's 2009 *World Population Data Sheet*.

There are ways to improve waste management and change behavior in developing countries, even if there is no budget for it. For instance, in Curitiba, Brazil, innovative waste collection approaches were developed, such as the "Green Exchange Program", to encourage slum dwellers to clean up their areas and improve public health. The city administration offered free bus tickets and fresh vegetables to people who collected garbage and brought waste to neighborhood centers. In addition, children in Curitiba were allowed to exchange recyclables for school supplies or toys.

Cities always need to find local solutions for waste management appropriate to their own particular circumstances and needs. In Delhi there is an army of over 1,20,000 informal waste collectors (so-called *Kabari*) in the streets, collecting paper, aluminum cans, glass, and plastic who sell the waste to mini-scrap dealers as part of a secondary raw materials market.

It is an informal industry which processes 59% of Delhi's waste and supports the livelihood of countless families. In the Indian capital city, the private sector does the waste management and the business of collecting and recycling is a serious one for many of the poor, and a relatively lucrative source of income. According to Bharati Chaturved, one out of every 100 residents in Delhi engages in waste recycling. Chaturved also estimated that a single piece of plastic increases 700% in value from start to finish in the recycling chain *before it is reprocessed*. This informal sector of waste collectors saves the city's three municipalities a large amount of costs of otherwise arranging waste collection, particularly in inaccessible slum areas. In Delhi, more than 95% of homes do not have formal garbage collection (Chaturved, 2010).

For countries like India or Bangladesh, the introduction of an industrialized clean-up system and perfected infrastructure like in the developed world would take jobs from thousands of poor peasants who are willing to work hard and get dirty collecting and recycling the waste of the metropolis in order to feed themselves. An estimated 6 million people in India earn their livelihood through waste recycling. On top of a low standard of living, they now face joblessness with India's new business model approach to waste management—replacing the preexisting informal Kabari system with a model from developed countries. It is an area where India and Bangladesh could probably learn from their neighbor China, since their cities have similar population densities (United Nations Human Settlements Programme, 2010).

Another interesting example for the informal waste management sector is the city of Cairo, the capital of Egypt, which has grown to over 15 million people and is one of the most densely populated cities in the world (with 32,000 people per sq mile). The economy of "Garbage City" (Manshiyat Naser, the Zabaleen quarter), a slum settlement on the outskirts of Cairo, revolves entirely around the collection and recycling of the city's garbage, mostly through the use of pigs by the city's minority Coptic Christian population. Although the area has streets, shops, and apartments, like any other area of the city, it lacks infrastructure and often has no running water, sewage or electricity. The city's garbage is brought in by the garbage collectors, who then sort through the garbage to retrieve any potentially useful or recyclable items. As a passer-by walks down the road he will see large rooms stacked with garbage, with men, women or children crouching and sorting the garbage into what is usable or what is sellable (Beitiks, 2010).

Families typically specialize in a particular type of garbage that they sort and sell—one room of children sorting out plastic bottles, while in the next room women separate cans from the rest. Anything that can somehow be reused or recycled is saved. Various recycled paper and glass products are made and sold from the city, while metal is sold by the kilogram to be melted down and reused. Carts pulled by horse or donkey are often stacked 3 m high with recyclable goods (see Figure 8.5).

The circular economic system in "Garbage City" is classified as an informal sector, where people do not just collect the trash, they live among it. Most families typically have worked for generations in the same area and type of waste specialization, and they continue to make enough money to support themselves. They collect and recycle the garbage which they pick up from apartments and homes in wealthier neighborhoods. This includes thousands of tons of organic waste, which is fed to the pigs. By raising the pigs, the Zabaleen people provide a service to those who eat pork in the predominantly Muslim country, while the pigs help to rid neighborhoods of tons of odorous waste that would otherwise accumulate on the streets. Like the famous "Smokey Mountain" rubbish dump in Manila, Philippines, could this place become an official recycling center?

As the cases in Delhi and Cairo illustrate, the increase in world flows of scrap, e-waste, recovered plastics and fibers has turned developed countries into a source of material supply for informal trade in emerging countries.

A global paradigm shift in urban development and the use of resources is essential. Clearly, a situation where 20% of the world's population consumes 80% of the world's resources cannot go on forever or be allowed to continue.

FIGURE 8.5   Many developing countries have such active informal sector recycling, reuse, and repair systems, which are achieving recycling rates comparable to those in developed countries, at no cost to the formal waste management sector, saving the city as much as 20% of its waste management budget. Cairo, for instance, has grown to over 15 million people and is one of the most densely populated cities in the world. The economy of "Garbage City" (Manshiyat Naser, the Zabaleen quarter), a slum settlement on the outskirts of Cairo, revolves entirely around the collection and recycling of the city's garbage, mostly through the use of pigs by the city's minority Coptic Christian population. Although the area has streets, shops, and apartments, like any other area of the city, it lacks infrastructure and often has no running water, sewage or electricity[12].

8.5.2   Composting Organic Waste and Improving Urban Ecology

Compost is an important source of plant nutrients and is a low-cost alternative to chemical fertilizers. It has become a necessary part of contemporary landscape management and urban farming, as it uses "reverse supply chain" principles, giving organic components back to the soil, thus improving the quality of agriculture. Paying attention to the nutrient cycle and to phosphorus replacement is part of sustainable urban agriculture. Industrial composting helps to improve soils. However, a proper composting infrastructure needs to be set up. The important focus on soil, putting nutrients back into agriculture (for instance, the "City to Soil" program in Australia). In Sweden, for instance, the dumping of organic waste to landfill has been illegal since 2005. It is essential to avoid landfill organics such as food waste. All organic waste should be used for composting or anaerobic digestion (see Figure 8.6).

Food waste is another major concern. 22% of all waste in Australia is food waste. New biodegradable packaging helps to facilitate processing of food waste. Biodegradable and compostable solutions for food waste recovery systems, using a kitchen caddy with a biodegradable bag that is collected weekly, has become a common solution. Iain Gull and, director of Zero Waste Scotland, points out that "over 60% of food waste is avoidable. However, if all unavoidable food waste in Scotland was processed by anaerobic digestion, it could produce enough electricity to run a city in size of Dundee" (Gulland, 2010). In South Australia more than 90,000 tons p.a. of food waste goes to landfill (on average, each household throws out 3 kg food waste per week). This needs to be taken out of the waste stream and diverted into composting or anaerobic digestion systems[13].

FIGURE 8.6   Photo: Organics recycling is important to return nutrients back to the soil, and there are new process improvements on a massive scale. Metropolitan green organics are collected through council curbside and industrial collections, as well as food organics (food scraps) from hotels, restaurants and supermarkets; composting and mulching transforms the material into a range of high-quality compost, mulch and soil products, to be returned to gardens and parklands[14].

8.6   CONCLUSION

8.6.1   Decoupling Waste Generation from Economic Growth

Because cities are the main consumers of energy, materials, food and water, it is essential that the delivery of urban services (including waste stream management and resource recovery) is as efficient as possible. The efficiency and effectiveness of urban services is greatly affected by the urban land-form (for instance, the low densities and mono-functional layout of suburbs is leading to highly inefficient conditions, often an increase in consumption and contributes to the problem).

Increased material and energy consumption in all nations, coupled with an inadequate and unsustainable waste management system, has forced governments, industry and individuals to put into practice new measures to achieve responsible, closed loop solutions in waste management and resource recovery. Achieving "zero waste" remains difficult and requires continued and combined efforts by industry, government bodies, university researchers and the people and organizations in our community.

The topic of reducing urban household consumption by optimizing urban form, and the need to reducing the material requirements for buildings (in fact, of the entire construction sector) has only recently emerged as an urgent field of further research (Dey et. al., 2007). While there is a general acknowledgment that there is a need for improved urban governance processes and rethinking of urban development patterns to reduce material consumption and optimize material flows, this is still a relatively new research field and there is still a lack of reliable data and comparative methodologies. One of the findings of this chapter is that embedding "zero waste" requires strong industry leadership, new policies and effective education curricula, as well as raising awareness (education) and refocusing research agendas to bring about attitudinal change and the reduction of wasteful consumption. Unlimited consumption and growth on a planet with limited resources "cannot go on forever and is indeed dangerous" (Meadows et. al., 1971).

The C&D sector has a particularly urgent need to catch up with other sectors in better managing its waste stream, to increase its focus on reusing entire building components at the end of a building's life-cycle. In Australia, for instance, around 40% of all waste to landfill comes from the building sector[15] (Recover Your Resources, 2009). Increasing the economic value of recycled commodities, such as rare metals in e-waste, paper, glass and plastics, remains an area for future development and investment.

Energy markets will soon compete with material markets for resources. The recycling sector in Germany employs already over 2,20,000 people in green jobs (2010). Waste is increasingly being seen in terms of economic sustainability, and it is a policy issue that offers great opportunities for the creation of green jobs.

A particular challenge in waste management is soil degradation. Composting methods are important to return nutrients from organics back to the soil. However, the anticipated global decline in the availability of phosphorous ("peak phosphors"), which is currently lost as waste from urban areas, however, is a vital nutrient for food production.

This chapter has touched on some of the complexities around sustainable urban metabolism, waste management and the links between waste streams, urban development, as well as the need for resource recovery. The three case studies are hopeful models of what could be achieved in Adelaide (Australia) and Aalborg (Denmark). These cases are of limited value for the developing world and large, rapidly expanding cities such as Delhi, Cairo and cities in China. Here, the informal sector of waste management deserves a closer look and more research focus. The import of waste to developing countries is obviously another interesting but complex issue: on one side, we criticize developed countries for their export of pollution, on another side; developed cities provide raw materials for workers in developing countries to mine urban waste. These informal sectors might even hold some lessons for cities in the developed world. Due to their greater consumption levels, cities in the developed countries have much higher material and energy consumption, despite the increase of resource efficiency[16] (Lenzen et. al., 2008; Rickwood et. al., 2008).

The developing world is fast catching up with consumption levels and will continue to increase its hunger for resources. China, for instance, is urbanizing faster than any other country ever before in history, requiring a huge amount of non-renewable materials, energy and water for the production of the consumer goods, and increasingly contributing to the depletion of raw material resources. The "new consumer" in Asia, who is part of a newly emerging middle-class, with resource-intensive lifestyle habits, materialistic behavior and mobility needs, contributes to and accelerates the development. Most of the consumption is going to be in cities. We can define a formula: The environmental impact (I) is a result of the increasing affluence/consumption power (A), a growing urban population (P) and the availability of technology (T). The suggested formula is: $I = P \times A \times T.$

It is essential that we continue to reduce wasteful consumption, to avoid the creation of waste in the first place (waste minimization through avoidance), to promote the cyclical reuse of materials in the economy and to maximize the value of our resources to make resource recovery common practice. Waste is a precious resource. The challenges posed by climate change and the depletion of resources are complex—but as a society we have the skills, knowledge and determination to achieve the necessary changes. Change to behavior, long-held planning habits and design attitudes will be necessary. In his latest book "A Final Warning", James Lovelock outlined the urgency and that time is critical[17,18,19] (Lovelock, 2009). In 2010, 6.8 billion people on the Earth consume resources, energy and materials in an ever increasing pace and volume. It is therefore essential to utilize 100% of all used resources as new resources, and embed the sustainable city paradigm, while drastically raising the efficiency of the use of resources, energy and materials (see diagram Figure 8.7).

In the meantime, nothing less than a peaceful revolution has started, changing the way we design, build, operate, maintain and recycle/renew cities and buildings. The urbanization process has emerged as the incubator and platform for revolutionary change: holistic strategies and integrated approaches for urban development indicate that *post-fossil fuel cities* can and must become the most environmentally-friendly model for inhabiting our earth. Waste avoidance has to be considered as one of the main drivers for architectural and urban design. In this context, our objective must be to reconcile the scarcity of our natural resources with the huge quantities of waste produced by our cities and industries, waste which we must, unfailingly, recover[20,21].

THE SUSTAINABLE CITY

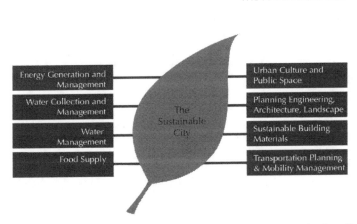

FIGURE 8.7   Diagram: Waste management is an important key stone in the effort towards achieving holistically a "Sustainable City"[22].

KEYWORDS

- Adaptive re-use of buildings
- Changing behavior
- Closed loop urban metabolism
- Material flow
- Product stewardship
- Recycling and reuse
- Reducing consumption
- Resource recovery
- Urban waste streams
- Waste avoidance
- Zero waste concept

ACKNOWLEDGMENT

The author wishes to thank the two anonymous reviewers and the editorial assistant for their helpful comments on earlier drafts of this chapter.

NOTES

[1] For further information on this author's work and research centre, see also: http://www.slab.com. au (accessed on Dec 1, 2010).

[2] A useful web site on re-using building components in architectural design is: http://www.superuse. org (accessed on Oct 30, 2010).

[3] See also this web site for more general informing about recycling: http://www.recyclicity.net (accessed on Oct 30, 2010).

[4] Quoted from: Recycling International (September, 2008). Available online: http://www.environmental-expert.com (accessed on March 1, 2010).

[5] More information at web site of Zero Waste SA government agency, Adelaide, Australia. http://www.zerowaste.sa.gov.au (accessed on October 30, 2010).

[6] On consumption of household products, see also this web site: http://www.thestoryofstuff.com (accessed on Oct 30, 2010).

[7] For more information on new production methods and material innovation, see: http://www.worldchanging.com (accessed on March 1, 2010).

[8] For more information on new production methods and material innovation, see also: http://www.transmaterial.net (accessed on March 1, 2010).

[9] For more information on new construction methods using timber systems, see: http://www.low-2no.org (accessed on Oct 30, 2010).

[10] Report: Study 2008. Institut der Deutschen Wirtschaft (IDW), Koeln/Bonn, Germany (2008). Available online: http://www.iwkoeln.de (accessed on Sept 1, 2010).

[11] More information on applied material flow management is available online at this web site: http://www.stoffstrom.org/en/institute (accessed on March 10, 2010).

[12] Photo: Courtesy Bas Princen, (2009). Available online: http://www.treehugger.com/files/2009/12/photographer-capture-life-in-garbage-city.php (accessed on 30 October 30, 2010).

[13] Information on food waste by Ronnie Kahn (Australia) on OZHarvest, the not-for-profit NGO, which delivered 6 million meals to people in need, between 2004 and 2010; and the "Love Food—Hate Waste" campaign. Available online: http://www.ozharvest.org (accessed on Oct 10, 2010). See also: www.lovefoodhatewaste.com (accessed on Oct 30, 2010). Information by John Dee (Australia) on food waste and better shopping methods. Available online: http://www.dosomething.net and www.foodwise.com.au (accessed on Oct 30, 2010).

[14] Photo: courtesy of the author, 2010.

[15] On resource recovery in the C/D sector, see also: http://www.resourcesnotwaste.org (accessed on Dec 20, 2010).

[16] Regarding recycled aggregates for concrete, see also: http://www.holcimforum.org (accessed on Oct 30, 2010).

[17] In terms of further (pessimistic) outlook and on the impact of global warming, see also: Brown, L. *Plan B 2.0: Rescuing a Planet under Stress and a Civilization in Trouble.* W.W. Norton Publishing, New York, USA (2006).

[18] For valuable data and evidence of European CO_2 emissions, see web site: http://www.euCO2.eu (accessed on Feb 1, 2010).

[19] Another useful web site with various data on environmental pollution and population growth: http://www.poodwaddle.com/worldclock.swf (accessed on Feb 1, 2010).

[20] More information on the German recycling system is available online: http://www.gruener-punkt.de/en/duales-system-deutschland-gmbh.html (accessed on Feb 1, 2010).

[21] Additional information on the German recycling system, which has set standards for best practice, see also: http://www.retech-germany.de (accessed on Feb 1, 2010).

[22] Diagram: courtesy of the author, 2008.

[23] © 2011 by the authors; licensee MDPI, Basel, Switzerland. This article is an open access article distributed under the terms and conditions of the Creative Commons Attribution license (http://creativecommons.org/licenses/by/3.0/).

REFERENCES

Barr, S. and Gilg, A. Sustainable lifestyles: Framing environmental action in and around the home. *Geoforum*, 37, 906–920 (2006).

Beitiks, M. *Incredible 'Garbage City' Rises Outside of Cairo*; Inhabitat (Nov 12, 2009). Available online: http://www.inhabitat.com/incredible-garbage-city-rises-outside-of-Cairo (accessed on Oct 30, 2010).

Blue: Water, Energy, and Waste, 29, Grimshaw Architects (Ed.). Andrew Whalley, London, UK, p. 56 (2010).

Braungart, M. and McDonough, W. *Cradle to Cradle. Remaking the Way We Make Things.* North Point Press, New York, USA (2002). *Note*: It has frequently been argued that the concept of "waste" should be substituted by the concept of "resource." For instance, Michael Braungart points out that the practice of dumping waste into landfill is a sign of a—failure to design recyclable, sustainable products and processes. In his research, Braungart focuses on flows of energy, water, materials, nutrients and waste. Process-integrated technology, as advocated by the "cradle-to-cradle" approach, includes the cascading use of resources in which high-grade flows are used in high-grade processes and residual waste flows are used in lower-grade processes, thus utilizing the initial value of a resource in the most efficient way.

Braungart, M. and McDonough, W. *Cradle to Cradle: Remaking the Way We Make Things.* North Point Press, New York, USA (2002). See also: MacKay, D. J. C. *Sustainable Energy—Without the Hot Air*; UIT Cambridge Ltd., Cambridge, UK (2009). In regard to learning from Nature, see also: Benyus, J. M. *Biomimicry: Innovation Inspired by Nature.* Harper Perennual, New York, USA (2002).

'Carrot City', *Building Design Proposal for Dallas with Integrated Urban Farming.* Project Illustration; courtesy MOV + Data, Lisbon, Portugal (March, 2009).

Chalmin, P. and Gaillochet, C. *From Waste to Resource. An Abstract of World Waste Survey, 2009.* Veolia/CyclOpen Research Institute, Paris, France (2009). *Note*: The "Gruene Punkt" (*green dot*) recycling system in Germany was introduced in 1991 and legislated in 1993, under the leadership of Minister Klaus Toepfer. One outcome is that today, 82% of all packaging is recycled. In Germany, economic growth has been decoupled from the amount of waste first time in 2008. At the end of its life-cycle, products are unlikely to end up on landfill or in incineration plants; over 87% of all aluminum and tinplate waste, as well as paper and plastic, is recycled and fully re-used in the production loop. Most paper manufacturers have also become paper recyclers.

Chaturved, B. Ragpickers: The bottom rung in the waste trade ladder. International Plastics Task Force (January, 2010). Available online: www.ecologycenter.org/iptf/Ragpickers/ indexragpicker.html (accessed on Oct 30, 2010).

Dey, C., Berger, C., Foran, B., Foran, M., Joske, R., Lenzen, M., Wood, R., and Birch, G. Household environmental pressure from consumption: An Australian Environmental Atlas. In *Water, Wind, Art and Debate: How Environmental Concerns Impact on Disciplinary Research.* G. Birch (Ed.). Sydney University Press, Sydney, Australia (2007).

Dey, C., Berger, C., Foran, B., Foran, M., Joske, R., Lenzen, M., and Wood, R. A Household environmental pressure from consumption: An Australian Environmental Atlas. In *Water, Wind, Art and Debate: How Environmental Concerns Impact on Disciplinary Research.* G. Birch (Ed.). Sydney University Press, Sydney, Australia, pp. 280–315 (2007).

Easton, S. Manufacturers to Arrange for Recycling of Australia's e-Waste. Available online: http://www.upiu.com/articles/manufacturers-to-arrangerecycling-for-australia-s-e-waste (accessed on Sept. 10, 2010).

Easton, S. *Manufacturers to Arrange for Recycling of Australia's E-Waste.* UPIU [Online], August (2010). Available online: http://www.upiu.com/articles/manufacturers-to-arrange-recycling -for-australia-s-e-waste (accessed on Sept 10, 2010).

Environmental Protection Agency (EPA). *Recover Your Resources: Reduce, Reuse and Recycle Construction and Demolition Materials at Land Revitalization Projects* (2009). Available online: http://epa.gov/brownfields/tools/cdbrochure.pdf (accessed on Oct 10, 2010). *Note*: The waste mix in Australia contains 22% food waste; 18% plastics; 22% glass; 12% paper; 6% cardboard. Each Australian produces 360 kg of organic waste per annum. Thereof, 250 kg are disposed off in landfill, significantly contributing to greenhouse gas emissions (Data: 2010). Australia produces approximate 45 million tonnes of waste annually, and around 50% of this (22 million tons) go to landfill (Data, Australian Waste Management Association: 2010). Australia currently operates 700 official, licensed landfills where methane gas from organic waste needs to be captured. However, organics in landfill are a major contributor to greenhouse gas emissions. The landfill levy varies widely from state to state, between $30 and $60 per ton. The recycling costs will always be more than landfill costs (except for steel, aluminum and other metals, which have high embodied energy and can easily be recycled). As material values of metals increase, the resource recovery for precious metals is likely to assist in increasing recycling rates.

Environmental Protection Agency (EPA). *Recover Your Resources: Reduce, Reuse and Recycle Construction and Demolition Materials at Land Revitalization Projects* (2009). Available online: http://epa.gov/brownfields/tools/cdbrochure.pdf (accessed on Sept 10, 2010). *Note*: Estimating that only 40% of the construction and demolition building materials were reused, recycled, or sent to waste-to-energy facilities and the remaining 60% was sent to landfills.

Faulstich, M. Technical University Munich, Munich, Germany. Personal communication (Sept 14, 2010).

Fischer-Kowalski, M. Society's metabolism. The intellectual history of material flow analysis, Part I. 1860–1970. *J. Ind. Ecol.*, 2, 61–78 (1998).

Girardet, H. *Cities, People, and Planet: Urban Development and Climate Change*, 2nd ed.; Wiley & Sons. Oxford, UK (2008).

Girardet, H. *Creating Sustainable Cities, (Schumacher Briefing #2)*. Green Books, Devon, UK (1999). Here, Herbert Girardet points out the importance for cities to adopt a circular metabolism: —In nature, waste materials are absorbed beneficially back into the local environment as nutrients. Cities don't do that. They work by way of taking resources from one place and dumping them somewhere else causing damage to nature. We need to turn this linear process into a circular process instead. The recycling of particularly organic waste is important for the sustainability of large cities. We need to meet this challenge and create a metabolism that mimics natural systems. Materials and products that we use need to be biodegradable. Plastic, which does not decompose easily, can be produced so that nature can absorb it more effectively.

Girardet, H. World Future Council, London, UK. Personal communication (Sept 14, 2010).

Grosse, F. Is recycling part of the solution? The role of recycling in an expanding society and a world of finite resources. *SAPIENS*, 3, 1–17 (2010). *Note*: The adaptive re-use of existing buildings is always a more sustainable strategy than building new. Instead of tearing down and rebuilding (which usually means losing the materials and embodied energy of the existing building), adaptive re-use allows the building to be given a new lease of life; an approach that was the norm until a generation ago. Now, our focus needs to return to upgrading the existing building stock. Recent research conducted by the Advisory Council on Historic Preservation (ACHP) in the U.S. indicates that even if 40% of the materials of demolished buildings are recycled, it would still take over 60 years for a green, energy-efficient new office building to recover the energy lost in demolishing an existing building.

Gulland, I. Zero Waste Scotland, Glasgow, UK. Personal communication (Sept 14, 2010).

Head, P. *Entering the Ecological Age*. Report on the Institution of Civil Engineers (ICE) Brunel International Lecture 2008, ICE, Arup, UK (2009). Available online: http://www.arup.com/Publications/Entering_the_Ecological_Age.aspx (accessed on March 10, 2010).

Hobson, K. Thinking habits into action: The role of knowledge and process in questioning household consumption practices. *Local Environment*, 8, 95–112 (2003). *Note*: Product stewardship refers to the responsible management of manufactured goods and materials, and includes regulated schemes (e.g., appliances, packaging and car take back schemes in the European Union) and industry-government agreements (e.g., *Australia's National Packaging Covenant*. Available online: http://www.packagingcovenant.org.au (accessed on Oct 10, 2010)).

Lane, R., Horne, R., and Bicknell, J. Routes of re-use of second-hand goods in Melbourne households. *Aust. Geographer*, 40, 151–168 (2009).

Lehmann, S. Sustainability on the Urban Scale: Green Urbanism. In *Urban Energy Transition*, P. Droege (Ed.). Elsevier, Amsterdam, The Netherlands (2008). In this chapter I point out that with dwindling raw materials and the depletion of resources, energy and material resources continue to be consumed and wasted at an accelerated rate. Our current energy generation method and distribution network has not changed much since the industrial revolution, and the way we extract virgin materials to consume them ("cradle to grave") has also not changed much either. The construction industry has a huge responsibility, as it is one of the most wasteful sectors. Concrete, for example, is responsible for a significant amount of global CO_2 emissions. For a while now, researchers have predicted that several construction materials will be exhausted by the end of the century.

Lehmann, S. *The Principles of Green Urbanism. Transforming the City for Sustainability.* Earthscan, London, UK, pp. 261–269 (2010).

Lenzen, M., Wood, R., and Foran, B. Direct *versus* embodied energy. The need for urban lifestyle transitions. In *Urban Energy Transition: From Fossil Fuels to Renewable Power.* P. Droege (Ed.). Elsevier, Amsterdam, The Netherlands, pp. 91–120 (2008). On the subject of recycled aggregates for concrete, see also: *Recycling Concrete: Cement Sustainability Initiative*; World Business Council for Sustainable Development, WBCSD, Geneva, Switzerland, pp. 1–8 (2009). Available online: http://www.wbcsd.org/includes/ (accessed on March 1, 2010).

Linacre, S. *Household Waste: Australian Social Trends.* Australian Bureau of Statistics, Canberra, Australia, p.10 (2007).

Lovelock, J. *A Final Warning,* Pinguin Books, London, UK (2009).

Meadows, D. H., Meadows, D. L., Randers, J., and Behrens, W. W., III. *The Limits to Growth,* Universe Books, New York, USA (1972).

Meadows, D. H., Meadows, D. L., Randers, J., and Behrens, W.W., III. The Clube of Rome. *The Limits to Growth,,* Universe Books, New York, USA , pp. 10–12 (1971).

On Poulsen and Hansen's Research, see also: Danish eco city proves waste management can reverse greenhouse trend. Science Daily (Nov 30, 2009). Available online: http://www.sciencedaily.com/releases/2009/11/091130103634.htm (accessed Oct 20, 2010).

On sustainable urban development, see also: United Nations World Commission on Environment and Development (WCED). *Our Common Future,* G. H. Brundtland (Ed.). Oxford University Press, Oxford, UK and New York, USA (1987).

Perkins, A., Hamnett, S., Pullen, S., Zito, R., and Trebilcock, D. Transport, housing and urban form: The life-cycle energy consumption and emissions of city centre apartments compared with suburban dwellings. *Urban Policy Res.,* **27**, 377–396 (2009).

Poulsen, T. and Hansen, J. A. Wastewater treatment and greenhouse gas emissions. *Waste Manage. Res.,* **27**, 861 (2009).

Public Review Landfill Capacity and Demand Report. see: [35] (2009). Full newspaper article, Waste Solution Left to Rot as Landfill Capacity Runs Out. *Sydney Morning Herald* (March 21, 2010). Available online: http://www.smh.com.au/environment/(accessed Dec 20, 2010).

Recover Your Resources: Reduce, Reuse and Recycle Construction and Demolition Materials at Land Revitalization Projects. Environmental Protection Agency (EPA), Washington, DC, SA (2009). Available online: wwwepa.gov/brownfields/tools/cdbrochure.pdf (accessed on Dec 20, 2010).

Rickwood, P., Glazebrook, G., and Searle, G. Urban structure and energy—a review. *Urban Policy Res.,* **26**, 57–81 (2008).

Rogers, R. and Power, A. *Cities for a Small Country.* Faber & Faber, London, UK (2000).

Satterthwaite, D. The implications of population growth and urbanization for climate change. *Environ. Urban.,* **21**, 545–567 (2009). Architecture and urban development must now address these new complex challenges that have arisen with our throw-away consumer society; architecture will need to deliver more, beyond the classical "practicality and beauty"; the entire process of building design, construction and product manufacturing must be based on future, more sustainable material flows and opportunities for resource recovery and R&D in the field of material efficiency and recyclability is increasingly important.

Schatzki, T. R. A new social ontology. *Phil. Soc. Sci.,* **33**, 174–202 (2003).

Shove, E., Watson, M., Hand, M., and Ingram, J. *The Design of Everyday Life.* Berg Publishing, Oxford, UK (2007).

South Australia's Draft Waste Strategy 2010–2015. *Consultation Draft*; Zero Waste SA: Adelaide, Australia (2010). Available online: http://www.zerowaste.sa.gov.au/upload/about-us/wastestrategy/DraftWasteStrategyV2.pdf (accessed on Sept 10, 2010). For instance, Australia's recent *National Waste Policy* subtitled "Less Waste—ess Wasteaste " (Australian Government, 2009) acknowledges both the growing waste problem and the need for research and action beyond current recycling initiatives to minimize the creation of waste. Furthermore, it emphasises the importance of product stewardship schemes that anticipate new roles and responsibilities for both producers and consumers of products and materials, see also: Australian Bureau of Statistics (ABS). *Australia's Environment Issues and Trends*; ABS Cat. No. 4613.0, Canberra, Australia (2006).

State Government of New South Wales. *Public Review Landfill Capacity and Demand Report,* Wright Corporate Strategy Pty Ltd, North Sydney, New South Wales, Australia (2009). Available online: http://www.planning.nsw.gov.au/LinkClick.aspx?fileticket=Xtm8jz6j7WI%3D&tab id=70& language=en-AU (accessed on Sept 10, 2010).

Tonkinwise, C. *De-materialism and the Art of Seeing Living. or Why Architecture's Self-Images lead to McMansions*. Presentation at the Faculty of Architecture, Design and Planning, University of Sydney, Sydney, Australia (2005).

United Nations Human Settlements Programme (UN-Habitat). *Solid Waste Management in the World's Cities: Water and Sanitation in the World's Cities, 2010*. Earthscan Ltd., London, UK, pp. 114–116 (2010).

United Nations Human Settlements Programme (UN-Habitat). *Solid Waste Management in the World's Cities: Water and Sanitation in the World's Cities, 2010*. Earthscan Ltd.: London, UK and Nairobi, Kenya (2010). Alone in Dhaka, Bangladesh, a quarter of a million children are living in the streets, surviving from the collection of garbage; they collect almost anything and sell it to micro scrap dealers, who sell it on to recycling companies.

Walls, M. *Extended Producer Responsibility and Product Design*. Report to Organization for Economic Cooperation and Development; Resources for the Future (RFF Press), Washington, DC, USA, pp. 69–71 (2006).

Waste Report 2010. SEA Organization: Riverside, CA, USA (2010). Available online: http://www.solv.org (accessed on Aug 10, 2010).

Weisenbach, D. Weisenbach Recycled Products (based in Columbus, Ohio, USA). Personal communication (Sept 14, 2010).

Williams, M. Cylch Corp., Wales, UK. Personal communication (Sept. 14, 2010).

World Population Growth Forecast. United Nations Human Settlements Program (UN-Habitat): Nairobi, Kenya (2010). *Note*: Global world populations in 2010 were 6.8 billion. It is predicted by UN-Habitat to increase to 9 billion by 2050. While the population in some countries is shrinking (Japan, Germany, Italy, Russia), other countries, such as India, have a fast growing population. The population in India is forecast to overtake that of China's by 2050 (India is predicted to have 1.6 billion people). We will soon reach the limits of the Earth's "carrying capacity" (what Rees and Wackernagel call "overshooting", 1996), for instance, the Earth's reduced capacity to supply fresh drinking water to all citizens of a city (as we have seen in Sub-Saharan African cities and in Mexico City). The world's population has been growing significantly since around 1800 due to the improved control of diseases and longer life expectancy. As a consequence, numerous scientists recommend halting further growth in cities in arid, hot climatic regions. At the same time, global agriculture is approaching a natural limit. While the amount of food production needs to keep increasing in pace with population growth, there is hardly any undeveloped farmland left on the planet. Experience shows that birth rates fall when women are well educated, when they aspire to a career, or when they chose to marry later and to have only one child. Clearly to slow down this immense population growth and to delay a food/water/energy supply disaster, we have to succeed in three important areas: reducing consumption and changing behavior; improving technology; and limiting population growth through education programs.

World Urbanization Prospects: The 2010 Revision. United Nations, Department of Economic and Social Affairs, Population Division: New York, USA (2010).

World Urbanization Prospects: The 2010 Revision. United Nations, Department of Economic and Social Affairs, Population Division, Nairobi, Kenya (2010).

ZWSA Survey (2009). Available online: http://www.zerowaste.sa.gov.au (accessed on Dec 10, 2010). The need to go beyond materials recycling is discussed: For durable household and electronic goods, increasing rates of obsolescence, ongoing toxicity issues and limited options for reuse or recycling are contributing to a growing waste problem. The scale of change required to reverse this trend cannot be achieved through recycling alone—as Linacre notes the increase in recycled goods has been countered by an increase in overall material flows. The inputs of resources and materials and outputs of wastes need to become more closely linked.

9 Integrated Solid Waste Management Based on the 3R Approach

M. A. Memon

CONTENTS

9.1 INTRODUCTION

Integrated solid waste management (ISWM) based on the 3R approach (reduce, reuse, and recycle) is aimed at optimizing the management of solid waste from all the waste-generating sectors (municipal, construction and demolition, industrial, urban agriculture, and healthcare facilities) and involving all the stakeholders (waste generators, service providers, regulators, government, and community/neighborhoods). This chapter discusses the concept of solid waste management (SWM). Initially, SWM was aimed at reducing the risks to public health, and later the environmental aspect also became an important focus of SWM. Recently, another dimension is becoming a critical factor for SWM, that is resource conservation and resource recovery. Hence, the 3R approach is becoming a guiding factor for SWM. On the one hand, 3R helps to minimize the amount of waste from generation to disposal, thus managing the waste more effectively and minimizing the public health and environmental risks associated with it. On the other hand, resource recovery is maximized at all stages of SWM. Lately, the new concept of ISWM has been introduced to streamline all the stages of waste management, that is source separation, collection and transportation, transfer stations and material recovery, treatment and resource recovery, and final disposal. It was originally targeted at municipal solid waste management (MSWM), but now the United Nations Environment Program (UNEP) is promoting this concept to cover all waste generating sectors to optimize the level of material and resource recovery for recycling as well as to improve the efficiency of waste management services. The ISWM concept is being transformed into ISWM systems to replace conventional SWM systems.

This chapter further discusses the implementation process for ISWM. The process includes a baseline study on the characterization and quantification of waste for all waste generating sectors within a city, assessment of current waste management systems and practices, target setting for ISWM, identification of issues of concern and suggestions from stakeholders, development of a draft ISWM plan, preparation of an implementation strategy, and establishment of a monitoring and feedback system. The UNEP is assisting member countries and their cities to develop an ISWM plan covering all the waste generating sectors within a specific geographical or administrative area such as a city or municipality. This umbrella approach is useful to generate sufficient volumes of recycling materials required to make recycling industries feasible. This is also helpful for efficient reallocation of resources for SWM such as collection vehicles, transfer stations, treatment plants, and disposal sites. The UNEP is assisting cities to develop and implement.

The ISWM based on the 3R approach. These experiences could be useful for other countries to develop and implement SWM to achieve improved public health, better environmental protection, and resource conservation and resource recovery.

The ISWM and 3R (reduce, reuse, and recycle) have become common terminologies for policy makers and practitioners in the field of SWM. However, in many countries ISMW is taken as being synonymous with traditional MSWM. In some countries, ISWM is understood to be an integrated approach for managing municipal waste to optimize the efficiency of services and to achieve the objectives of the 3R approach.

This chapter discusses the concept of ISWM and argues that ISWM may go beyond municipal waste management alone and may cover all the waste generating sectors to optimize the efficiency of the services at each stage of waste management and to increase the amount of recoverable materials and energy to make it attractive for the private sector. Stages of the ISWM chain include source separation, collection and transportation, transfer stations and material recovery, treatment and resource recovery, and final disposal. Waste management services include the technology and human resources to facilitate the flow of waste and recovery at each stage. Furthermore, it is suggested that 3R is inherently integrated within ISWM.

This chapter also highlights the process of developing and implementing ISWM in cities/towns. This process includes establishing baseline waste data and assessment of current waste management systems, target setting, identification of stakeholders' issues of concern for ISWM, and development of an ISWM plan with its implementation strategy.

9.2   EVOLVING CONCEPT

The ISWM is an evolving concept. Initially ISWM was developed to increase the efficiency of the MSWM chain, that is source separation, collection and transportation, transfer stations, treatment, and final disposal. (Tchobanoglous et al., 1993) Later, ISWM became an umbrella management system to coordinate all waste types from all waste sources (residential, commercial, industrial, healthcare, construction and demolition, and agriculture) within a geographic or administrative boundary such as a city. Furthermore, ISWM became a process to achieve 3R, aiming to minimize the quantity of waste requiring disposal and to maximize recovery of material and energy from waste. Thus, ISWM is a system based on the 3R approach at the city/town level covering all waste generating sectors and all stages of the waste management chain, including segregation at source for reuse and recycling, collection and transportation, sorting for material recovery, treatment and resource recovery, and final disposal.

9.2.1   Background

The ISWM started evolving right from the beginning. Historically, solid waste was considered as the waste produced by humans and animals, consuming resources to support life. (Tchobanoglous et al.,1993) Later, with industrial activities, the scope of solid waste was broadened to include the wastes generated by industry. Later, it was also realized that catastrophic events such as earthquakes, floods, and fire also generate debris. This debris, the result of natural disasters or the outcome of construction and demolition activity, is also considered to be solid waste that needs to be removed. The management of solid waste was not a major issue when the population was small and the land available for the assimilation of wastes was large. (Tchobanoglous et al.,1993)

Furthermore, the impact of waste on public health was not yet fully realized. However, after the outbreak of the worst public health impacts, especially in Europe, the removal of waste became one of the top priorities for public health. This was not only applicable to biodegradable wastes, which produce disease-related vectors, but was also applicable to non biodegradable wastes, which were accumulating and resulting in urban flooding and were affecting sanitary conditions.

The initial success of maintaining public health by removing waste from cities and dumping it outside did not last for long because open dumps and open burning started having its own impact on public health and on the natural environment. Leach ate from dumps started seeping into water resources and into agricultural fields, resulting in contamination of water and food. Local air pollution from burning of waste increased the incidence of various diseases.

This led the public and governments to give serious thought to the proper management of solid waste so that it would not affect public health and the natural environment directly or indirectly. The SWM became a priority public service for local governments. At this time, SWM services were mainly considered for municipal solid waste (MSW); thus, MSWM was a common term with varying definitions in different parts of the world. Hester and Harrison indicate that depending on the country, the definition of MSW can include some or all household wastes, including hazardous wastes; bulky wastes; street sweepings and litter; parks and garden wastes; and wastes from institutions, commercial establishments, and offices. (Hester and Harrison, 2002)

Industrial waste management became the responsibility of waste generators (industries) as well as national governments. In countries with increased decentralization such as Japan and China, local governments were also responsible for regulating and monitoring industrial waste management.

Since then, new types of waste have emerged, such as wastes from healthcare services, wastes from discarded electronic equipment including computers (e-waste), waste from end-of-life vehicles (ELV), wastes from urban agriculture, and huge waste quantities from construction and demolition activities and from catastrophic events such as urban floods and earthquakes.

9.2.2   Responses to Managing Waste

Waste management is one of the costliest public services. Conventional responses to collection, transportation, treatment, and disposal of waste in an environmental friendly way became a burden due to the rapid increase in waste generation levels as a result of urbanization and economic growth. Developing countries are in the worst situation because most modern waste collection, treatment, and disposal equipment is imported and the revenue base to support waste management is very small. Table 9.1 and Figure 9.1 show the expenditures on MSWM by selected countries and cities. The financial burden started to become critical with an increase in energy and land prices.

TABLE 9.1   Expenditures on municipal solid waste management (MSWM) (From MacFarlane, 1998)

City. country	Year	Per capita expenditure on MSWM (US$)	Per capita GNP (US$)	Percentage of GNP spent on MSWM
New York, USA	1991	106	22240	0.48
Toronto, Canada	1991	67	20440	0.33
Strasburg, France	1995	63	24990	0.25
London, UK	1991	46	16550	0.28
Kula Lumpur, Malaysia	1994	15.25	4000	0.38
Budapest, Hungary	1445	13.80	4130	0.33
Sao Paulo, Brazil	1989	13.32	2540	0.52
Buenos Aries, Argentina	1989	10.15	2160	0.47
Tillinn, Estonia	1995	8.11	3080	0.26
Bogota, Columbia	1994	7.75	1620	0.48
Caracas, Venezuela	1989	6.67	2450	0.27
Riga, Latvia	1995	6	2420	0.25
Manila, Philippines	1995	4 (estimated)	1070	0.37
Buchanrest, Romania	1995	2.37	1450	0.16
Hanoi, Vietnam	1994	2 (predicted)	250	0.80
Madras, India	1995	1.77	350	0.45

TABLE 9.1   *(Continued)*

City. country	Year	Per capita expenditure on MSWM (US$)	Per capita GNP (US$)	Percentage of GNP spent on MSWM
Hanoi, Vietnam	1994	2 (predicted)	250	0.80
Madras, India	1995	1.77	350	0.45
Lahore, Pakistan	1985	1.77	390	0.45
Dhaka, Bangladesh	1995	1.46	270	0.54
Accra. Ghana	1994	0.66	390	0.17

GNP. gross national product

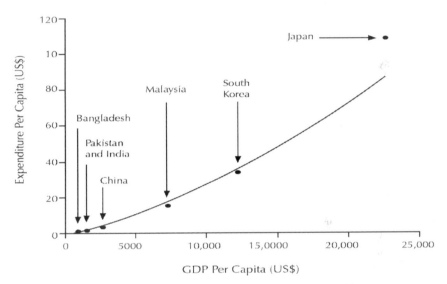

FIGURE 9.1   Examples of national expenditure levels on municipal solid waste management. (From MacFarlane, 1998).

The waste collection rates in many developing countries were affected badly due to rapid increases in the cost. It became very difficult to find land near a town for landfilling, and transportation costs and environmental impacts became major constraints to constructing landfills at a distant location.

Hence, the most vital response was to reduce the amount of waste. Reduced quantities of waste would decrease the burden on collection services as well as on treatment and final disposal facilities. Various strategies, including technological and policy based, were introduced to reduce the amount of waste at the point of generation. Cleaner production (CP) is being introduced to minimize the waste generation by industry, while awareness-raising campaigns and waste collection fees were introduced to motivate residents, institutions, commercial entities, and others to limit their waste generation levels.

9.2.3   Current State of Waste Generation

Local, national, and international efforts were made to raise the awareness of waste generators to reduce the amount of waste generation. There were substantial gains, especially for controlling the levels of industrial waste generation. However, municipal waste was still on the rise, and it is estimated that in 2004 the total amount of MSW generated globally reached 1.84 billion tonnes, a 7% increase over 2003.(Global Waste Management Market Report, 2004; Memon and Matsuoka, 2002) It is further estimated that between 2004 and 2008, global generation of municipal waste will rise by 31.1%, roughly a 7% increase annually. New emerging waste streams, especially with hazardous waste components are also arising. The Secretariat for the Basel Convention (SBC) estimated that about 318 and 338 million tonnes of hazardous and other waste was generated in 2000 and 2001, respectively, (Memon and Matsuoka, 2001) based on incomplete reports from the parties to the Convention. Healthcare waste is classified as a subcategory of hazardous waste in many countries. The World Health Organization (WHO) estimates that in most low-income countries, total healthcare waste per person per year is anywhere from 0.5 to 3 kg. (Memon et al., 2006) There is no comprehensive estimate about global industrial waste generation. The Environmental Protection Agency of the United States of America (US EPA) estimates that American industrial facilities generate and dispose of approximately 7.6 billion tonnes of nonhazardous industrial solid waste each year. (Memon et al., 2005) Waste from agriculture and rural areas includes both biomass agricultural residues and hazardous wastes such as spent pesticides. The European Union (EU) estimated that its 25 member states produce 700 million tons of agricultural waste annually. (Memon et al., 2003)

TABLE 9.2   Waste management practices

Region	Sanitary land-fill (%)	Incineration	Open dumps (%)	Recycling (%)	Open burning (%)	Others (%)
Africa	29.3	1.4	47.0	3.9	9.2	8.4
Asia	30.9	4.7	50.0	8.5	1.7	4.5
Europe	27.6	13.8	33.0	10.7	11.8	4.4
North America	91.1	0.0	0.0	8.1	0.0	0.0
Latin America	60.5	2.0	34.0	3.2	5.5	2.0

9.2.4   Current State of Waste Management in Developing Countries

The World Bank estimates that in developing countries, it is common for municipalities to spend 20–50% of their available budget on SWM, and still 30–60% of all urban solid waste is uncollected and less than 50% of the population is served. In most developing countries, open dumping with open burning is the norm. (Kochi et al., 2001) In low-income countries, collection alone uses up 80–90% of the MSWM budget. In middle-income countries, collection costs 50–80% of the total budget. In high-income countries, collection accounts for less than 10% of the budget, which allows large funds to be allocated to waste treatment facilities. Upfront community participation in these advanced countries reduces the collection cost and facilitates waste recycling and recovery. Despite various efforts and community-based initiatives, the overall situation of waste management remains challenging, as shown in Table 9.2.

9.2.5   Concept of Integrated Solid Waste Management Based on 3R

The scenario discussed in the preceding section reflects the challenges of conventional integrated waste management, which was sector specific and had little emphasis on resource recovery for reuse and recycling. The major challenge was that most of the funds were being consumed by collection of waste and it was almost impossible for many countries to support proper treatment and disposal without external funding.

The international agencies realized that improvements in waste management could not be achieved through a piecemeal approach. An integrated approach was required to reduce the increasing amount of waste that requires proper collection, treatment, and disposal. However, efforts to minimize waste through awareness-raising and policy could result in substantial reductions in volumes of waste. In addition to that, it was also realized that waste contains precious resources that could be recovered in terms of materials for recycling as well as in terms of energy to be used as a substitute for fossil fuels. This realization completes the concept of 3R to reduce the final amount of waste as well as to divert most of the waste for reuse and resource recovery. The reduced amounts of waste could substantially decrease the costs for waste management. Resource augmentation by converting waste into material or energy could broaden the revenue base to support expenditures for SWM.

Initially, this 3R approach was promoted in each waste sector individually, mainly due to the institutional framework in most countries where local government is responsible for municipal waste and construction and demolition waste, and national government is responsible for industrial waste and agricultural waste. However, it was realized that by integrating various sectors under the ISWM concept of umbrella management, there would be various gains. First, the available resources for waste collection, material recovery, treatment and resource recovery, and disposal could be used efficiently with better scheduling and higher resource use efficiency. Second, there would be substantial amounts of recovered materials and energy available to facilitate the establishment of industries that could use these resources for production. Third, there would be savings in waste management costs as the overall amount of final waste that requires disposal would be reduced considerably through diversion of waste for material and resource recovery. Fourth, there would be active coordination among various stakeholders that could lead them to work on other development projects such as water and sanitation. Fifth, the outcome of ISWM in terms of cleaner and safer neighborhoods would lead to improved quality of life, better economic activity, and higher property values. Last, but not least, governments can build trust among the public as ISWM brings tangible outcomes in terms of public health, jobs and economic gains from recycling industry, cleanliness, and active interactions among stakeholders. Hence, the ISWM concept can optimize the gains of 3R on one hand, and improve the waste management system on the other hand. Figure 9.2 captures the ISWM concept based on the 3R approach.

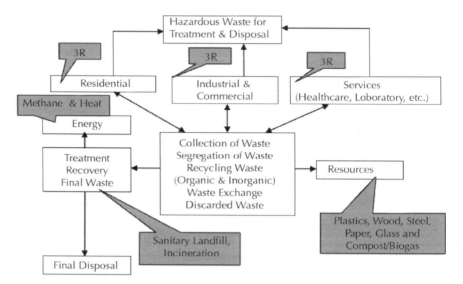

FIGURE 9.2 Concept of integrated solid waste management (ISWM) based on the 3R approach. (From UNEP-DTIE-IETC, 2009).

9.3 IMPLEMENTING ISWM

An ISWM system based on the 3R approach can be optimally designed and implemented at the town/city level due to the basic role of local government in providing waste collection and management services. However, the regional/provincial and national governments have to play very important roles, especially in terms of enacting appropriate policies and regulations as well as strengthening the institutions to create an enabling environment for ISWM.

Traditionally, many cities in developing countries did not have a dedicated waste management plan and waste management had a low priority for most local and national governments. In many cities, waste management was considered as the collection of garbage and the dumping of that garbage outside the city. Even for waste collection, a systematic approach was not adopted as the operational plan and the number of collection trucks was not designed based on waste generation rates. There is a clear difference in the new ISWM approach that requires a logical system based on reliable baseline data to cover collection as well as all the other stages of the waste management chain. Hence the designing and implementation of ISWM for a given city requires various steps, involving all the major stakeholders. These steps include:

1. Data collection and analysis to develop baseline data on the characterization and quantification of waste from various sources and future projections.
2. Information collection and analysis to develop baseline data on the current waste management system and gaps therein.
3. Setting of targets by local government in consensus with local stakeholders for ISWM.

4. Identification of issues of concern of local stakeholders covering financial, technical, environmental, and social aspects of ISWM.
5. Development of an ISWM plan.
6. Development of an implementation strategy for ISWM.
7. Development of a monitoring and feedback system for ISWM.

9.3.1 Waste Characterization and Quantification

To prepare an ISWM plan, the most important step is to collect substantial and accurate information on the quantity of waste and its composition as well as to project waste data for future scenarios. For waste characterization and quantification, primary data collection is essential. However, prior to starting the collection of primary data, proper groundwork should be carried out. This includes defining the administrative and geographical boundaries of the targeted city, identifying the waste generating sectors within the city, collecting maps showing zoning, and collecting basic information regarding city and secondary data if available (Figure 9.3). Based on this groundwork, a proper plan for data collection and analysis should be formulated and resources, including human resources and equipment, should be organized.

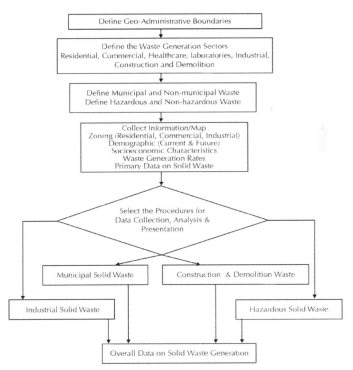

FIGURE 9.3 Process for data collection and analysis. (From UNEP-DTIE-IETC, 2009)

The most important decision for data collection is the number, location, and timing of samples. Sample collection and analysis is a costly activity, hence, excessive data collection and analysis could jeopardize the resources allocated for this activity. Samples could be collected at the generation level, at transfer stations, or at disposal sites. This depends on the coverage of the existing collection system. If all the waste is collected and transferred properly, then samples at transfer stations and/or disposal sites may provide reliable information. Sometimes, data collection is completed within a few months and this may not capture seasonal variations in waste quantity and characterization. In this case, based on local knowledge, adjustments could be made and these adjustments should be verified during the following season by collecting and analyzing representative samples.

The samples collected at the generation level or at transfer/disposal sites should be analyzed to ascertain the quantity of overall waste as well as quantities from each source and at each district/street level.

The different components, including biodegradable (kitchen and yard) waste, plastic, paper, textiles, glass, metals, and others, should also be quantified for designing material and resource recovery systems. Waste samples should also be analyzed to assess the moisture content and calorific value to assist identification of appropriate technologies for collection, treatment, and disposal. Furthermore, based on relevant factors such as population growth and economic development, projections of waste quantities and changes in waste composition should be calculated for the future. A time period of 30 years, divided into 5 year sub periods, could be very helpful in designing an ISWM system and related infrastructure. All this information would be compiled to develop a baseline report on waste data.

This set of detailed data on waste quantity and characterization, with projections for the future, is essential to design an ISWM system (policies, technologies/infrastructure, financial mechanisms, and roles and responsibilities of stakeholders) to promote 3R. Some policies and technologies could be applied upstream, before the generation of waste, to minimize waste generation. However, most policies, technologies, and roles are targeted to promote reuse and recycling of waste through source separation, material recovery at transfer stations, and resource recovery at treatment centers. This will reduce the amount of waste to optimize the waste collection, transportation, and disposal system.

9.3.2 Assessment of Waste Management Systems

The second part of a baseline study would be the assessment of current waste management system/practices and identification of gaps therein. Waste management systems include the policies, institutions, technologies and infrastructure, financing mechanism, stakeholders' roles, and operational plan for waste collection (Figure 8.4). Policies for waste management cover local and national policies and the level of enforcement. The regulations as well as fiscal policies for SWM should be assessed to identify the gaps therein, either in policies or in enforcement. Assessment of institutions would provide information on the type and level of institutions involved with management of solid waste from one or more sources.

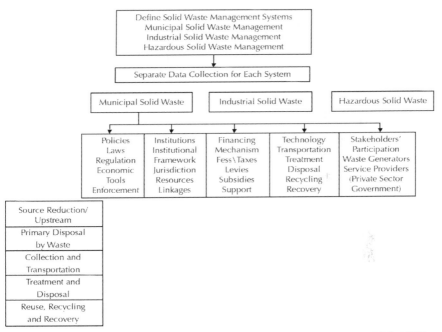

FIGURE 9.4 Assessment of waste management system. (From UNEP-DTIE-IETC, 2009)

This also helps to identify the shortcomings of current institutional arrangements with respect to efficiency and effectiveness of the SWM system. Assessment of technologies and infrastructure covers collection types (e.g., door-to-door, kerbside), type of collection vehicles, transfer station and sorting technology, treatment plants (e.g., incineration, composting/biogas), and landfill. The operational plan for waste collection, transfer stations, and disposal should also be analyzed. The operational plan goes beyond preparing a list of the technologies and infrastructure. For example, for the waste collection system, the number and type of waste collection vehicles and their status (operational and nonoperational) is one issue, but the operational plan, that is how the service provider or government operates these vehicles on a daily or weekly basis and how much waste is collected, is another issue. Similarly, the operational plan for transfer stations and landfill sites should also be analyzed.

Thereafter, the financial mechanism to support SWM in general and services (collection, treatment, and disposal) in particular should be analyzed to identify any gaps in revenue generation and expenditure. This financial mechanism may comprise local and national government support, international cooperation, and direct taxes or fees for the waste generators. It is also useful to conduct interviews of waste generators to assess their willingness to pay (Memon and Matsuoka 2002) or use the benefit transfer function (Memon and Matsuoka 2001) to assess the willingness to pay if there is limited time and resources for primary data collection.

Assessment of the stakeholders' role should cover all the major stakeholders such as waste generators, service providers, and regulators. Gap analysis should also be carried out to assess any shortcomings in the current SWM system. These shortcomings could be identified from two viewpoints: the traditional viewpoint of waste collection, treatment, and disposal and the new viewpoint of 3R, focusing on source separation, material recovery at transfer/disposal sites, and resource recovery at treatment centers.

9.3.3 Target Setting

Once the baseline report is ready, the next step is to set the quantitative targets for ISWM. These targets should be verifiable for monitoring and feedback. The target setting is led by local government and by involving all the major stakeholders, including waste generators, service providers, and the community as a whole. The targets should be in line with the "mission" and "vision" statements of a city or a country, if available. Otherwise, the starting point could be local or national goals. These statements or goals may include keywords such as clean city, public health, resource augmentation or 3R, and environmental friendly practices. The targets may cover all the stages and services with respect to ISWM. For example, targets for segregation at source may identify types of waste, such as food waste, to be segregated at source and the percentage and amount of waste to be segregated. For collection, the efficiency targets could be set, such as 100% collection of the waste generated. For material recovery, targets may be set for the amount of waste to be sorted to recover recyclable materials such paper, plastic, and metals. Similarly, targets for treatment may cover biological and thermal treatment and recovery of resources such as compost, biogas, ethanol, and heat/electricity. Finally, the targets for safe disposal may cover the volumes of hazardous and nonhazardous waste for controlled and sanitary landfill and recovery of landfill gas as a source of energy.

In addition to the targets for each stage of ISWM, other related targets such as broadening of the revenue base, increasing stakeholders' participation, or promoting public–private partnerships could also be included. Moreover, it is important to set the timeline for the targets. For example 90 and 100% collection efficiency of waste generated within the city should be achieved by the year 20XX and the year 20XY, respectively.

9.3.4 Identification of Stakeholders' Concerns

Based on the baseline report and proposed targets for ISWM, the stakeholders may have concerns and suggestions. These concerns could be categorized under financial, technological, environmental, and social issues. These concerns should be in line with the ISWM system based on the 3R approach. Hence, all the stakeholders should be briefed on the current situation of waste generation and waste management and on the implications of the 3R-based ISWM system. The major stakeholders would be waste generators, service providers, regulators, government, formal and informal sectors related to recycling, and community organizations. They may have concerns relating to one or more stages of ISWM, that is source segregation, collection, transportation, sorting and material recovery, treatment and resource recovery, and final disposal (Figure 8.5).

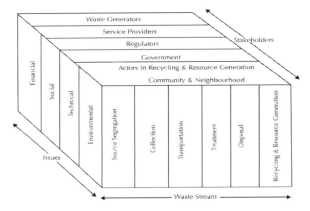

FIGURE 9.5 Issues of concern to stakeholders. (From UNEP-DTIE-IETC, 2009)

9.3.5 Management System

Once the baseline report is ready, targets have been proposed, and issues of concern have been identified, the next step would be to develop a management system to achieve the proposed targets for ISWM based on the 3R approach (Figure 8.6). Such a management system is a set of responses in terms of regulations/policies and institutional frameworks, technologies and infrastructure, and voluntary actions for each stage of ISWM (Figure 9.7). The policies, including regulatory and fiscal, and their enforcement, as well as the role of institutions, could be proposed in line with the existing system as ISWM is an evolutionary process and not a revolutionary process. It should be remembered that in many countries, national governments are responsible for policies; thus, practical policies with a proper timeline for their approval should be proposed.

Technological and infrastructure measures could be very costly; thus, based on the local socioeconomic situation and local capacity to operate and maintain these technologies, appropriate technologies should be proposed.

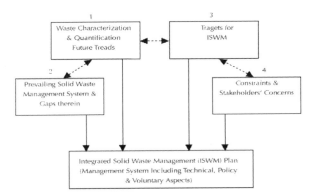

FIGURE 9.6 Developing ISWM. (From UNEP-DTIE-IETC, 2009)

Appropriate technologies could be identified by setting up criteria covering financial, technical, environmental, and social indicators. There are various frameworks available to assess the technologies. The UNEP Division of Technology, Industry, and Economics (DTIE), International Environmental Centre (IETC) has developed a sustainability assessment of technologies (SAT) framework to assist decision makers in selection of appropriate technologies for any targeted public service. Figure 9.8 shows an SAT framework-based analysis of a hypothetical example for selection of appropriate treatment and disposal methods. The criteria for SAT are developed based on the local economic, social, technical, and cultural conditions. Points are allocated for each criterion with relevant importance. For example, if the number of jobs is more important than the cost of the technology, then "number of jobs" may be evaluated from 1 to 10 points for each competing technology and "cost" may be evaluated from 1 to 5.

Last, but not least, some of the responses could be proposed as "voluntary" to make these responses popular among the stakeholders, and over time, these could be transformed into policies. Japanese experiences of pollution management through voluntary actions (Kochi et al. 2001; Memon et al., 2005, 2003, 2006) could be taken into consideration when setting up voluntary responses under the ISW

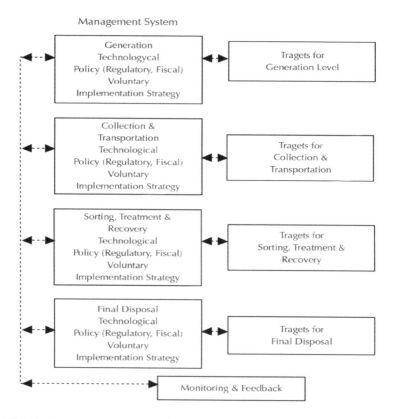

FIGURE 9.7   Management systems under ISWM. (From UNEP-DTIE-IETC, 2009)

9.3.6 ISWM Plan

An ISWM plan is a document containing baseline information, proposed targets, issues of concern and a set of responses as a management system, an implementation strategy, and monitoring and feedback systems (Figure 9.9). The implementation strategy for ISWM defi nes the ways and means of implementing each response. For a policy-level response, for example to propose an incentive on source separation in terms of a tax rebate, a proper strategy based on local conditions should be formulated. In some countries, local governments are sufficiently decentralized to take these decisions, while in other countries, national governments alone can take such decisions. Furthermore, for the policy-level response, the implementation strategy for

ISWM should also cover local capacity building to implement such policies, once these are approved at the appropriate level of government. For example, to increase waste collection coverage up to 100% of waste generated from all sectors within a city, a certain number of collection vehicles are required to be procured and put into proper operations. In the implementation strategy document, a detailed plan should be formulated to get the funding from possible sources, procedures should be put in place to procure the vehicles, and an operational plan should be implemented to operate and maintain the vehicles. Various investment and management strategies, such as public–private partnerships for the collection system, could also be the part of this document.

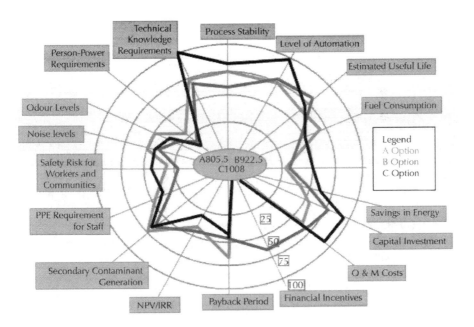

FIGURE 9.8 Sustainability assessme nt of technologies (SAT) framework based assessment of technologies: Star diagram. (From UNEP-DTIE-IETC, 2009)

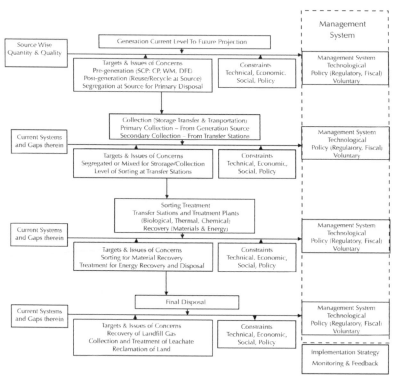

FIGURE 9.9   ISWM plan. (From UNEP-DTIE-IETC, 2009)

9.4   CONCLUSION

This chapter helps to understand the evolving concept of ISWM, based on the 3R approach, and the process for implementing an ISWM system in cities/towns. The ISWM system clearly improves resource use efficiency, as all waste sources are managed under an integrated waste management system. This is crucial for managing special wastes, such as hazardous waste. If individual sectors are managed separately, then it would be a costly business. Hence, applying joint efforts under ISWM could be efficient and effective. This is a major challenge for cities. Furthermore, resource recovery from one sector, such as the commercial sector, may not be adequate to attract investments in ecoindustries or to convert waste into a resource. Therefore, addressing all sectors under ISWM could be a very effective tool to manage their waste effectively and efficiently based on the 3R concept. Implementation of ISWM is straightforward because local capacity building, supported by national and international initiatives, can lead to all the actions being undertaken locally, including waste characterization and quantification, assessment of the current waste management system, targets for ISWM, identification of stakeholders' issues of concern, and development of an ISWM plan and implementation strategy for ISWM.

KEYWORDS

- **Assessment and planning**
- **Solid waste**
- **3R Approach**
- **Waste management**

REFERENCES

Global Waste Management Market Report (2004).

Hester, R. E. and Harrison, R. M. (Eds.). *Environmental and health impact of solid waste management activities: Issues in environmental science and technology.* The Royal Society of Chemistry, Manchester (2002).

Kochi, I., Matsuoka, S., Memon, M. A., and Shirakawa, S. Cost–benefit analysis of the sulfur dioxide emissions control policy in Japan. *Environ. Econ. Policy. Stud.*, **4**(4), 219–233 (2001).

MacFarlane, C. *What a waste: Solid waste management in Asia.* World Bank, Washington, DC (1998).

MacFarlane, C. In World Bank, *What a waste: Solid waste management in Asia,* Washington DC: 1999 (1998).

Memon, M. A., Imura, H., and Hitsumoto, R. Urban environmental management: Local capacity building through international cooperation. *Environ. Syst.*, **31**, 245–256 (2003).

Memon, M. A., Imura, H., and Shirakawa, H. Reforms for managing urban environmental infrastructure and services in Asia. *J. Environ. Dev.* **15**(2), 138–157 (2006).

Memon, M. A. and Matsuoka, S. Benefit transfer function to estimate WTP for rural water supply in Pakistan. *J. Int. Dev. Stud.*, **10**(2), 101–119 (2001).

Memon, M. A. and Matsuoka, S. Validity of contingent valuation estimates from developing countries: Scope sensitivity analysis. *Environ. Econ. Policy. Stud.*, **5**(1), 39–61 (2002).

Memon, M. A., Pearson, C., and Imura, H. Inter-city environmental cooperation: The case of the Kitakyushu Initiative for a Clean Environment. *Int. Rev. Environ. Strat.*, **5**(2), 531–540 (2005).

Tchobanoglous, G., Theisen, H., and Vigil, S. A. *Integrated solid waste management: Engineering principles and management issues.* McGraw-Hill, Singapore (1993).

UNEP-DTIE-IETC guidelines for data collection and analysis (2009).

UNEP-DTIE-IETC guidelines for assessment of waste management system (2009).

UNEP-DTIE-IETC guidelines for identification of issues of concern (2009).

UNEP-DTIE-IETC guidelines for ISWM Pla9n (2009).

UNEP-DTIE-IETC sustainability assessment of technologies (SAT) framework (2009).

10 Assessment of the Status of MSWM in Kamareddy Municipal Council

Syeda Azeem Unnisa* and S. Bhupatthi Rav

CONTENTS

10.1   INTRODUCTION

Design and implementation of sustainable municipal solid waste management systems (SMSWS) is a real challenge for developing countries. This is particularly so in places with very high urbanization rates and very low public awareness. Any management strategy in this sector will be sustainable only if it is cost effective, environmentally friendly, and implemented with the active and continuous involvement and participation of the public. This paper presents an assessment of the existing situation of municipal solid waste management (MSWM) in Kamareddy Municipality belonging to Nizamabad District in the State of Andhra Pradesh, India. Efforts to improve the systems are outlined. The chapter concludes with specific approaches for Sustainable MSWM for the Kamareddy.

The increase in the population in class I cities is very high as compared to that in class II cities. In India out of the total population of 1,027 million as on 1st March, 2001, about 742 million live in rural areas and 285 million in urban areas. The percentage of urban population to the total population of the country stands at 27.8. The net addition of population in rural areas during 1991–2001 has been to the tune of 113 million while in urban areas it is 68 million (Bolaane, 2006). The percentage decadal growth of population in rural and urban areas during the decade is 17.9 and 31.2% respectively (NEERI Report, 2005). By 2011 and 2021, the urban population is likely to be increased by 81 million and 174 million respectively. Thus, there has been an increase of 2.1% points in the proportion of urban population in the country during 1991–2001 (Petts, 2001). The uncontrolled growth in urban areas has left many Indian cities deficient in infrastructure services such as water supply, sewerage, and MSWM (Refsgaard and Magnussen, 2008). Urban sanitation and environmental issues are clearly contributors to basic health conditions in urban areas but Municipal Solid Waste (MSW) Management has a lower priority than water supply and sanitation (Bhoyar et al., 1996). The study was taken with the following objectives: assessment of waste quantity, assessment of existing status of collection, storage, transportation, treatment, and disposal activities. Suggestions for indicative strategies and guidelines enabling the municipal authorities to formulate an action plan for better management of MSW.

10.2 MSW (MANAGEMENT AND HANDLING) RULES, 2000

According to MSW Rules, 2000, every municipal authority is responsible for setting up a waste processing and disposal facility, and for preparing an annual report. The State governments and Union Territory Administrations have the overall responsibility for enforcement of the provisions of these rules in the metropolitan cities and within territorial limits of their jurisdiction (Municipal Solid Waste Rules, 2000).

10.3 STUDY AREA

Kamareddy is a large town located in the district of Nizambad in the state of Andhra Pradesh and located at 18°–17′ 29″ North Latitude and 78°–19′ 11″ East Longitude is the Southern Part of Nizamabad District. It is located NH-7 It is situated at a distance of 111 km from Hyderabad city. Kamareddy is a fast growing 2nd Grade Municipality having spread over 14.11 sq.km. The population of the town is 64,496 as per 2001 census. Kamareddy Town was constituted into a Municipality vide GOMS No. 743 Dt. 24-08-1987 as 2nd Grade under A.P. Municipalities Act, 1965. The land use is mostly for residential purposes. However, trading centers, educational institutions, and commercial activities bring in a large floating population. The details of population growth and about the municipality are given in Table 10.1 and Table 10.2.

TABLE 10.1 Population growth for every 10 years

Sl. No.	Year	Population	Variation in Population	Percentage
1.	1941	5282	--	--
2.	1951	7829	2548	48.22%
3.	1961	10318	2489	31.79%
4.	1971	17835	7517	72.85%
5.	1981	33048	15213	85.30%
6.	1991	48666	15618	47.26%
7.	2001	64496	15830	36.93%

TABLE 10.2   Kamareddy Municipality at a glance

1	Population(as per 2001 censes		64496
2	Area (in Sq.kms)		14.11
3	No.of Households		19500
4		(a) Kutcha Drains	40
	Length of Drains (in kms)	(b) Pucca Drains	110
		(c) Strom Water Drains	20
5		(a) CC Roads	50
	Length of Roads (in kms)	(b) BT Roads	25
		(c) WBM Roads	25
		(d) Kucha Roads	80
6	No. of Wards		33 Wards
7	No. of Slums	(a) Notified	17
		(b) Non-Notified	2
8	No. of Slaughter Houses	(a) Sheep & Goat	Nil
		(b) Beef	Nil
9	No.of Markets	Vegetable Market	1
10	No. of Educational Institutes	(a) Schools	25
		(b) Colleges	10
11	No. of Hospitals	(a) Nursing Homes	06
		(b) Govt. Hospitals	02
12	No. of Function Halls		10
13	No. of Hotels		12
14	No. of Cinema Halls		06
15	Railway Station		01
16	Bus depot		01
17	Bus Stands		02
18	No. of Hostels Boys and Girls		10
19	No. of Toilets	(a) Public	03
		(b) Private	Nil
20	Toilets comparative to Households		Nil
21		Households, Markets and Commercialestablishments	33 MT
	Daily Garbage Generation	(a) Slaughter houses	Nil
		(b) Silt and Debris	4.5 MT
22	Land for Compost Yard	(a) Land available (acres)	7 acres

10.4   METHODOLOGY

In the present study, Kamareddy Municipality was with the aim of covering all the SWM activities like door to door collection, segregated storage of wastes at the source, segregated collection, hygienic handling and transportation of wastes, and adoption of appropriate methods of disposal. The major activities included performing field investigations to assess the quantity of MSW generation per day and determining waste composition and characteristics. In addition, requisite secondary data were collected from the municipal authorities using a predesigned questionnaire.

10.5   ANALYSIS OF EXISTING PRACTICES OF MSWM IN KAMAREDDY

The detail study report prepared by the Regional Centre for Urban and Environmental Studies, Hyderabad puts the total quantum of solid waste generated in the city at 37.5 MT/day, resulting in an estimated per capita solid waste generation of 800 g/day. Prima facie, it appears to be on the higher side. Kamareddy estimates a floating population of 1.5 lakhs/day, bringing down the per capita waste generation close to 600 g/day. The major source of generation of solid wastes is the domestic sector with about 70,000 households. Other sources include commercial centers, hospitals, markets, construction and so on. Vegetable markets generate solid wastes in the range of about 40–50 tons/day, the construction industry around 20 tons/day, and silt from drains contributes about 7.5 tons/day. Figures available with the RCUCES, Hyderabad indicate that about 35 MT of wastes is being collected and transported to the dump yard, where processing is not yet started. According to the survey analysis, the composition of solid wastes is as follows: Biodegradable wastes—70%, Recyclable wastes (paper, plastic, metal, rubber, glass etc)—15%, Inert wastes—10%, Others—5%. The C:N ratio is put at around 40, which is higher than the ideal ration for composting. Moisture content is estimated to be around 55% and calorific value in the range 800–1,100 Kcal/kg.

An important milestone of this municipality in the implementation of SWM was the intensive awareness campaign carried out for various stakeholders followed by execution of a segregated system of storage of wastes at the source and segregated collection and hygienic handling of wastes. An action plan was chalked out for organizing primary as well as secondary collection. Women volunteers (Self Help Groups (SHGs) of women from BPL families) were engaged for house-to-house collection of solid wastes. A group of 10 members were formed in each ward for organizing primary collection. In some wards, two groups were formed, based on economic feasibility. The CRPs (Community Resource Persons) were imparted training to help them efficiently organize the primary collection. Bins with lids are provided in the auto-rickshaws so that the waste collected from households can be transferred to it. Primary collection of wastes is done in the morning from 6 a.m. Secondary collection and transportation is carried out in covered tractor-trailers. These have two compartments so as to enable collecting wastes in a segregated manner. Vehicles for secondary collection are so routed that the wastes from the auto-rickshaws can be directly transferred into them and secondary collection points are avoided.

To start with, it was decided to segregate solid wastes into two, viz. the biodegradable and the non-biodegradable fraction. In order to ensure and promote segregated storage of wastes at the source, Kamareddy Municipality distributed two bins to each household—one white (for storing the non-biodegradable fraction) and the other green (for storing the biodegradable fraction). Further, the Corporation directed all commercial establishments and hospitals to install bins on their own for this purpose. A debris service was introduced for removing construction waste and debris on payment. A system for collecting waste from gardens once a week was also launched. Dumper containers are provided only in the markets and slums and post box type bins at bus stops and busy junctions to prevent littering. A few months after implementation, it was found that in urban areas where the population density is very high, the scheme was working reasonably well. However, in the outer and relatively less thickly populated areas, participation was only about 50% because people with land preferred to dispose of their wastes on their own

Regular sweeping of streets, on daily basis, is carried out only on the main roads and certain parts of municipality centre. Some streets are swept alternate days and others once a week. It is estimated about 50% of roads are covered daily or alternately. Sweepers use normal brooms and baskets to sweep, collect the waste in small heaps, and this is then removed by handcarts. The waste so collected is transferred to box type containers or wheelbarrows and subsequently to an open ground-level secondary storage point. Workers are engaged for sweeping, drain cleaning and silt collection.

Although, a processing plant is proposed at Siricilla Road in 2009 it is not functional. The non-biodegradable rejects were just left in the open space within the plant area. The plant could not be operated efficiently due to various problems such as lack of roofed space for storing wastes without being open to rain, absence of adequate space for disposing the rejects and non-biodegradable fraction, lack of segregation yard, absence of platform and proper leachate collection and treatment facilities, roads, and drains and so on. A major problem in rehabilitation and modernization of the disposal plant was the huge quantum of waste that had piled up over the years and rejects from wastes that had already been processed. The piled up waste was not segregated and this made its handling complicated. Also, this waste lay in the open and it was not possible to process this waste for almost 6 months a year due to the very high moisture levels following rains. The existing land is 8 acres for treatment and disposal of waste which is 4 km away from the town.

10.6   PROBLEMS IN IMPLEMENTATION SWM

After a flying start, the municipality has presently run into trouble. A major issue in waste collection is the resistance to change, at least in some sections of the society, creating hurdles in adopting the concept of segregated storage and collection of wastes. In spite of the awareness drive, segregation at source is not very satisfactory. It appears that training and awareness programs for all stakeholders have to be conducted for some more time to ensure sustainability of the project. It has been reported that the door-to-door collection of biodegradable wastes, which was originally scheduled to be carried out daily, is now being carried out only once in 3 days or so in many residential colonies. About one-third of the auto-rickshaws, which were provided to the MEPMA CRPs or DWAKWA workers units for primary collection of wastes, are under repair. Also, it is reported that there has been a drop in the number of MEPMA CRPs or DWAKWA workers volunteers in recent times. Unless this is resolved urgently, the plant may come to a standstill. Other issues include low coverage by MEPMA CRPs or DWAKWA workers units, high operating cost of autos, open and exposed secondary storage in major collection points and markets, and manual handling for transfer to transportation vehicles.

The practices adopted by the Kamareddy Municipality for collection, transportation, and disposal of MSWs were totally unsatisfactory. The municipality had installed a few community storage bins and containers at selected locations. Households and establishments deposited wastes in these bins. However, these were grossly inadequate. The wastes were subsequently collected manually in auto rickshaws and transported to the Dump yard at Siricilla Road, located about 4 km away from Kamareddy. A number of rag pickers made their living by collecting a variety of recyclable wastes from bins and the disposal site. Although a processing plant had been proposed at the disposal site, its functioning is not yet started. Some major deficiencies of the system were as follows: (i) no segregation at the source, (ii) poor waste collection system at the primary level, (iii) uncontrolled littering, (iv) open community bins of very small capacities,

(v) using drains as solid waste disposal sites, (vi) mixing hospital wastes, (vii) open burning of wastes, (viii) poor waste collection system, (ix) lack of a notified collection schedule, (x) weak transportation system, (xi) unscientific and inappropriate processing scheme, (xii) lack of community participation, and (xiii) lack of trained staff.

10.7 STRATEGIES FOR SUSTAINABLE MUNICIPAL SOLID WASTE MANAGEMENT

- Segregated storage and, collection of solid wastes (wherever necessary), is an essential step towards sustainable solid waste management and hence should be practiced in urban areas.
- Wherever possible, household management and segregation of biodegradable waste should be encouraged and adopted.
- Whether a centralized solid waste disposal system is required or not is a question that has to be addressed with utmost care and caution. In Kamareddy, this may not be needed at all, as biodegradable wastes can mostly be handled at household level. Wherever this is not possible, efforts should be made by local bodies to utilize available space for solid waste disposal jointly. This can solve problems like financial difficulties currently faced by plant operators in Kamareddy.
- In Kamareddy, a scientific solid waste disposal facility is yet to be operational, so it would be ideal to investigate the feasibility of having decentralized units for various zones. This strategy may be needed, as land is fast becoming a scarce commodity in most towns.
- Currently, there is a lot of misconception among the administrators as well as the public on "landfilling". A vast majority from these groups believes that landfill is just a covered dump. This is far from the truth. Wherever a landfill is required to be established, it shall be designed, constructed, and operated as an "engineered landfill".
- In view of the improved living standards of the people, leading to the increasing use of different types of electronic goods, the amount of e-waste is expected to increase considerably in the coming few years. Local bodies have not made any attempt to quantify this hazardous waste. It is high time that they initiate moves to assess the quantum of e-wastes likely to be generated, so that this could push for an appropriate strategy to handle these at the State level.
- The responsibility for safely disposing hazardous and e-wastes shall be fixed on the manufacturer and a system of effective "take back" shall be implemented.
- Although considerable efforts have been made to build awareness among stakeholders on the importance and necessity for adopting environmentally sound, techno-economically feasible, and socially acceptable solid waste management practices, it is generally observed that results from the field are far from satisfactory. Hence there is an urgent need to intensify extension activities so as to continuously motivate and educate the stakeholders. Simultaneously, the principle that "the polluter pays" has to be strictly implemented.
- Existing legislation in this sector shall be strictly enforced.

10.8 CONCLUSION

Study carried out in the Kamareddy Municipality has revealed that there are many shortcomings in the existing practices followed for the management of MSW. These pertain mainly to inadequate manpower, financial resources, and implements/machinery required to effectively carrying out various activities of MSWM. In most of the cities, the waste quantity is not measured and is usually assessed based on number of trips made by transportation vehicles.

Proper records for timely action are not maintained. Based on the data collected and the assessment carried out, it is necessary to initiate improvement measures. To overcome the deficiencies in the existing MSWM systems, an indicative action plan incorporating strategies and guidelines should be delineated. Based on this plan, municipal agencies can prepare specific action plans for their respective ULBs. A need also exists to strengthen existing monitoring mechanisms, particularly from the point of view of implementation of provisions made in MSW (Management and Handling) Rules, 2000.

KEYWORDS

- **Biodegradable**
- **Disposal**
- **Management**
- **Solid Waste**
- **Sustainable Practices**

REFERENCES

Bhoyar, R. V., Titus, S. K., Bhide, A. D., and Khanna, P. Municipal and Solid Waste Management in India. *Indian Assoc. Environ. Manage.*, **23**, 53–64 (1996).

Municipal Solid Waste (Management and Handling) Rules. Government of India, New Delhi (2000).

Bolaane, B. Constraints to promoting people centered approaches in recycling. *Habitat Int.*, **30**, 731–740 (2006).

NEERI Report. Assessment of Status of Municipal Solid Waste Management in Metro Cities, State Capitals, Class I Cities and Class II Towns (2005).

Petts, J. Evaluating the effectiveness of deliberative processes: Waste management case-studies. *J. Environ. Planning Manage.*, **44**, 202–226 (2001).

Refsgaard, K. and Magnussen, K. Initial(s) Household behavior and attitudes with respect to recycling food waste-experience from focus group. *J. Environ. Manage.*, **12**, 1–12 (2008).

Index